Gender, Sexuality and R
in Evolutionary Narrative

Since the early 1990s, evolutionary psychology has produced widely popular visions of modern men and women as driven by their prehistoric genes. In *Gender, Sexuality and Reproduction in Evolutionary Narratives*, Venla Oikkonen explores the rhetorical appeal of evolutionary psychology by viewing it as part of the Darwinian narrative tradition.

Refusing to start from the position of dismissing evolutionary psychology as reactionary or scientifically invalid, the book examines evolutionary psychologists' investments in such contested concepts as teleology and variation. The book traces the emergence of evolutionary psychological narratives of gender, sexuality and reproduction, encompassing:

- Charles Darwin's understanding of transformation and sexual difference;
- Edward O. Wilson's evolutionary mythology and the evolution–creationism controversy;
- Richard Dawkins' molecular agency and new imaging technologies;
- the connections between adultery, infertility and homosexuality in adaptationist thought.

Through popular, literary and scientific texts, the book identifies both the imaginative potential and the structural weaknesses in evolutionary narratives, opening them up for feminist and queer revision. This book will be of interest to students and scholars of the humanities and social sciences, particularly in gender studies, cultural studies, literature, sexualities, and science and technology studies.

Venla Oikkonen is an Academy of Finland Postdoctoral Researcher in Gender Studies at the University of Helsinki. Her research interests include literature and science, evolutionary theory, population genomics, and scientific discourses of gender, sexuality, race and nation.

Transformations: Thinking Through Feminism

Edited by:
Maureen McNeil, Institute of Women's Studies, Lancaster University
Lynne Pearce, Department of English, Lancaster University

Other books in the series include:

Gender, Sexuality and Reproduction in Evolutionary Narratives

Venla Oikkonen

LONDON AND NEW YORK

First published 2013 by Routledge

2 Park Square, Milton Park, Abingdon, Oxon OX14 4RN
711 Third Avenue, New York, NY 10017, USA

Routledge is an imprint of the Taylor & Francis Group, an informa business

First issued in paperback 2017

British Library Cataloguing in Publication Data
A catalogue record for this book is available from the British Library

Library of Congress Cataloging in Publication Data
Oikkonen, Venla.
  Gender, sexuality and reproduction in evolutionary
narratives / Venla Oikkonen.
    p. cm. -- (Transformations : thinking through feminism)
  Includes bibliographical references and index.
1. Evolutionary psychology. 2. Human evolution. 3. Human reproduction.
4. Sex differences. I. Title.
  BF698.95.O55 2013
  155.3--dc23
            2012043934

ISBN: 978-0-415-63599-8 (hbk)
ISBN: 978-1-138-09468-0 (pbk)

Typeset in Times New Roman
by Taylor & Francis Books

Contents

Acknowledgments

This book would not have been possible without the generous support of several individuals, communities, and institutions. I thank the University of Helsinki, the Finnish Research School in Women's and Gender Studies, the Kone Foundation, and the Academy of Finland for providing the financial means that enabled me to give my full attention to this project over several years.

The research for this book was conducted at the University of Helsinki in 2004–12. I have been very fortunate to be a member of the gender studies community in Helsinki, first as a doctoral student, then as a lecturer, and most recently as an Academy of Finland postdoctoral researcher. I thank my PhD advisors Kirsi Saarikangas, Leena-Maija Rossi, and Nely Keinänen, who provided invaluable suggestions and encouragement especially in the early stages of the project. My colleagues in gender studies in Helsinki have created a generous, supportive, ambitious, unconventional, and truly interdisciplinary academic home. I thank especially Maija Urponen, Anne Soronen, and Eeva Urrio for sharing their thoughts and comments on issues that turned out to be formative for the book, Nina Järviö and Outi Pajala for practical assistance on numerous occasions, and Aino-Maija Hiltunen, Eva Maria Korsisaari, and Hanna Johansson for solidarity and support. I thank Johanna Kantola and Marjaana Jauhola for practical advice and comradeship in the final stages of the book project, Tuija Pulkkinen for her encouragement, and my wonderful officemates Anna Moring, Minna Seikkula, Mervi Patosalmi and Anna Elomäki for cheering me through the last hectic weeks of revisions.

I wish to thank Maureen McNeil at Lancaster University and Priscilla Wald at Duke University, whose comments and encouragement have helped me situate my work within the larger context of feminist cultural studies of science. I thank Sabine Sielke for the opportunity to present and discuss the book project at the University of Bonn in February 2012. In Finland, my collaboration with Sari Irni and Mianna Meskus has provided an inspiring dialogue that has pushed me to rethink my theoretical premises and methodological choices. I also wish to thank my colleagues in the Academy of Finland project *Representing and Sensing Nature, Landscape and Gender*, in which I participated in 2007–10, as well as the other members of the national gender studies doctoral program, in which I participated in 2005–6.

On a more personal note, I want to thank my family and friends, who have provided all those things in life that make a project like this possible in the first place. Particularly heartfelt thanks go to Joseph Flanagan, on whose unfailing support I could always count. Our myriad conversations on an endless range of topics—cultural, political, academic—have kept me thinking. Our son Liam was born halfway through this project. Liam has showed me the wonderful richness of life and the importance of everyday events, experiences, and emotions.

Two chapters of this book are based on previously published articles. An earlier and shorter version of Chapter 3, entitled "Narrating descent: popular science, evolutionary theory and gender politics," was published in *Science as Culture*, volume 18, issue 1 (2009). I thank Taylor and Francis for the permission to include a revised version of the article here. A shorter version of Chapter 4, entitled "Mutations of romance: evolution, infidelity, and narrative," was published in *Modern Fiction Studies*, volume 56, issue 3 (2010). I thank the Johns Hopkins University Press for the permission to include the work in this book.

Introduction

During the opening week of the 1998 World Cup in France, the British news magazine *The Economist* published an editorial titled "Sex, death and football" (*The Economist* 1998). The text suggests that human males' assumed addiction to football is underwritten by "the Darwinian rationale," thus placing the athletic event in the context of evolutionary explanations of gender and sexuality (*The Economist* 1998: 18). The text portrays male behavior humorously through the language of reversed evolution, as it predicts that "men all over the planet will regress to traditional patterns of behavior" and ignore their families and domestic responsibilities (*The Economist* 1998: 18). Women, the editorial insists, are not likely to behave this way: although women may appear to be interested in football, they are actually interested in the male footballers or the male fans. The text invokes a reported link between life expectancy in primate species and the time spent taking care of offspring, lamenting "the terrible price that men pay for their devotion, and the shocking biological advantage that women derive from indulging it" by monopolizing childcare (*The Economist* 1998: 18). The editorial closes with advice for men to "[c]hange a nappy, by God, and put years on your life" (*The Economist* 1998: 18).

Ten years later, in 2008, the electronics company Braun launched an advertising campaign for a new electric razor. The campaign included three ads, each of which placed an image of a primate—a chimpanzee, a baboon, and a gorilla—side by side with an image of a human male. In each ad, the primate face is captioned as "8:00 am" and the human face as "8:05 am" while the text underneath reads: "Braun Series 1—Brings out the human in men." This parallelism posits evolution as the basis of (male) human existence. It also suggests that modern masculinity is a product of repeated daily performance, as men are still essentially animals—a representation that resonates with the popular idea of male sexuality as driven by uncontrollable animal desires. This rhetoric is not new in advertising: for example, the awarded 2005 Guinness "noitulovE" commercial "rewinds" evolution from three men sipping Guinness in a pub to three lizard-like creatures by a prehistoric pond, thereby rendering evolution a teleological procession toward idealized Western masculinity.

This book could have opened with hundreds of other examples of evolutionary discourse, widely available in magazines, newspapers, science news,
documentaries, popular science, prime-time television, films, fiction, and dinner-
party conversations. Since the emergence of evolutionary psychology in the
early 1990s in particular, evolutionary arguments have achieved considerable
prominence in contemporary culture. Evolutionary psychology is the study of
human cognition and behavior as a product of evolutionary processes. The
approach builds on the theoretical framework of sociobiology, introduced in
the 1970s, in which animal behavior is explored within its evolutionary context. Evolutionary psychology gained momentum in the 1990s through its
popular association with the Human Genome Project, the biotechnological
enterprise that sequenced and mapped the overall of human genetic material.
The project endorsed the view of the genome as an evolutionary archive,
thereby offering a powerful rhetoric that could produce culturally appealing
claims about the prehistoric roots of human nature. As Dorothy Nelkin and
M. Susan Lindee put it: "Biology, in a very real sense, has become a philosophical and religious domain, and the genome itself has become a guide to
the human condition" (Nelkin and Lindee [1996] 2004: xvii).

With the entanglement of human genomics and evolutionary psychology
in popular discourse, gender and sexuality have emerged as the prime locus
of the debate about human behavior. As a result, claims about evolution and
genomics tend to be interpreted as having implications for our understanding
of gender and sexuality, while gender and sexuality, even in nonscientific
contexts, are often assumed to have an evolutionary foundation. The genetically informed evolutionary discourse that permeates the public sphere often
represents evolutionary psychology as the synonym of evolutionary theory.
As modern evolutionary biology includes numerous disciplines that do not
address human behavior, or address it from a viewpoint radically different from
that of evolutionary psychology, this association is both curious and troubling.
Moreover, the evolutionary psychological discourse that we encounter in the
media is highly problematic. For example, it appeals to personal experiences of
gendered phenomena like football, or invokes *a priori* assumptions about nature
and knowledge, such as "the Darwinian rationale" in *The Economist*. These
assumptions often stand as a rhetorical substitute for actual evidence in popular
discourse. They are also reinforced through unsubstantiated reference to other
biological disciplines, as in the invocations of primatology in *The Economist*
editorial and the Braun commercial. These moves across disciplines and
discourses reduce highly complex behaviors and notoriously fickle desires to
fantasized moments of human origins in the prehistoric savanna.

Not surprisingly, the rhetorical appeal of evolutionary discourse has alarmed
academics in the humanities and social sciences.[1] In particular, the enormous
popularity of evolutionary psychological arguments about gender and sexuality
has engendered a sense of urgency among scholars working in gender and
cultural studies. At the heart of this growing concern is the concept of change:
while feminist, queer, and cultural studies scholars work on the premise that

culture is—and should be—mutable, evolutionary psychologists tend to embrace the idea of modern genders and desires as molded in the Pleistocene. The question that critiques of evolutionary psychology need to address, then, is how to challenge such politics of immutability. This is the question that I seek to answer in this book.

This book argues that recent evolutionary arguments can be effectively understood only in the context of the narrative and discursive history from which they sprang. That is, by viewing currently circulating evolutionary narratives as mutations of earlier Darwinian narratives—including that of Darwin himself—we may identify the cultural and structural conditions of evolutionary narration. The chapters that follow trace transformations of evolutionary narrative from Darwin to sociobiology to evolutionary psychology. These travels will take us across genres, debates, and historical contexts. Through these acts of textual border crossing I wish to show how narrative analysis is able to locate and dissect the organizing logic of evolutionary narration. Most importantly, narrative analysis allows us to identify discursive fractures, inherent contradictions, and structural weaknesses in the seemingly infallible logic of immutability that underlies today's evolutionary discourse.

From morality to promiscuity

The politics of immutability and change that I trace in this book is closely connected to the longstanding debates about the biological underpinnings of sexual identities and behaviors such as monogamy, polygamy, or homosexuality. All these issues are at the heart of scientific controversies over human evolution. While the roots of the debate lie in Charles Darwin's texts and extend even further in the history of evolutionary thought, there is a clear discursive shift between Darwin's and later biologists' understanding of the nature of desire. In order to set the scene for the investigation of the evolutionary politics of transformation, stability and foundationality, I begin with a brief overview of this discursive shift.

Charles Darwin's famous exploration of the place of humanity in nature, *The Descent of Man*, was first published in 1871. The book argued that one of the most significant events in human evolution had been the emergence of "the moral sense or conscience" (Darwin [1879] 2004: 120). While the rudiments of moral behavior were present in many animal species, the human species had acquired its capacity for true morality through the cultivation of the social instinct necessary for communal living and the consequent feeling of sympathy for other members of the community. Like his contemporaries, Darwin understood the evolution of morality as having culminated in the Western nations, which supposedly demonstrated the highest level of self-command, the prerequisite of ethical reflection. Among the presumably less developed races, by contrast, animal instincts still ruled with the result that "[u]tter licentiousness, and unnatural crimes, prevail to an astounding extent" (Darwin [1879] 2004: 143), as evident in "the profligacy of the women" in primitive societies.[2] For Darwin, this gap in attitudes toward "the self-regarding virtues" (Darwin [1879] 2004: 157) between primitive and civilized societies suggested that "the hatred of indecency, which appears to us so natural as to be thought innate" is in fact "a modern virtue, appertaining exclusively ... to civilized life" (Darwin [1879] 2004: 143). The ability to resist carnal and corporeal urges, then, emerges in *The Descent* as the final stage in the narrative of human progress. At the same time, the origin of that success story is firmly located in the realm of nature.

While heatedly debated at the time of publication, Darwin's ideas of human evolution and the role of sexuality received relatively little attention in the first decades of the twentieth century. In the 1970s, however, human evolution became the subject of a new scientific controversy. In 1975, Harvard ethologist Edward O. Wilson published his contested classic *Sociobiology: The New Synthesis*, which advocated the study of human behavior as a product of Darwinian evolution under the new field of sociobiology. Wilson's efforts were echoed in other widely read and equally controversial works such as Richard Dawkins' *The Selfish Gene* (1976) and David P. Barash's *The Whisperings Within: Evolution and the Origin of Human Nature* (1979). These texts all emphasized the fixity of gendered behavior and the key role of sexuality in evolutionary processes.

While sociobiology claimed to continue the work began by Darwin in *The Descent*, there was a significant change in the rhetoric employed in portraying modern humanity. In sociobiology, the moral autonomy celebrated by Darwin emerged as a mere illusion. Instead, our acts were driven by our "hidden masters," the scheming, self-absorbed genes (Wilson 1978: 4), which "whispered within" us (Barash 1979), "build[ing] brains in such a way that they tend to gamble correctly" and thus "propagate those same genes" (Dawkins [1976] 1999: 56). Within such a framework, moral behavior appeared as a disguised reproductive strategy "programmed to a substantial degree by natural selection" (Wilson 1978: 6). Represented by "the basic unit of selfishness," the Dawkinsian gene (Dawkins [1976] 1999: 36), opportunism replaced altruism, encouraging individuals to "increase their Darwinian fitness through the manipulation of society" (Wilson [1975] 1982: 548). This discursive change engendered narratives of human evolution in which humans were given only static roles while their genes—the true evolutionary protagonists—pursued, competed, and conquered.

This framing generated an abundance of narratives that portrayed promiscuous and self-interested behaviors as evolutionary strategies. However, these sociobiological accounts typically excluded women from the realm of sexual liberties. As the gender burdened by pregnancy and lactation, women, the argument goes, could not engage in equally licentious behaviors. This representation of male sexuality as the driving force of evolutionary change was challenged by feminist sociobiologists—perhaps most notably Sarah Hrdy (1981)—who extended the logic of sexual opportunism to the females of the species. The sociobiologically inspired evolutionary narratives that circulate in contemporary culture often build on one of these narrative models. What they all share is a naturalized understanding of sexuality as the engine of the narrative of human evolution. This narrative engine is understood as engendering change, futurity, and potentially progress. Let us take a closer look at this narrative logic by examining its current manifestations.

"It's all about sex"

In 2003, the BBC produced and broadcast an ambitious documentary series, *Walking with Cavemen*,[3] which told the story of human evolution with the help of cutting-edge prosthetics and special effects. The opening episode developed an emotionally appealing (and highly hypothetical) narrative around its central character, "Lucy," the *Australopithecus afarensis* fossil found in Ethiopia in 1974 and named after the enigmatic heroine of the Beatles' "Lucy in the Sky with Diamonds." Lucy's story is represented as an account of how our hominid ancestors' "lives have led to the amazing reality of ours," a narrative framing that emphasizes adaptations—in Lucy's case, bipedalism—understood as crucial steps on the evolutionary path toward modern humanity. Halfway through the episode, however, the discussion of the evolution of bipedalism takes a wonderfully unexpected turn as Robert Winston, our host on this scientific expedition, turns to the camera and explains: "The truth is, walking on two legs has become a defining feature of my life and yours for the most surprising of reasons. Ultimately, it's all about sex." The fact that this statement is uttered in a confidential tone by an amiable middle-aged professor fond of picnicking on cookies in the African savanna produces a somewhat comical effect. But what is most striking about the scene is that, apart from the brief introduction of the two males competing for Lucy's attention, what precedes and follows this statement lacks any explicit reference to actual sex. After the 30-minute account of Lucy's life, the viewer is still left wondering what exactly the connection between sex and bipedalism might be.

That such a generously budgeted and carefully scripted production as *Walking with Cavemen* so light-handedly invokes sex as the ultimate explanation of evolutionary change suggests that sexuality is commonly understood as a key part of the narrative of human evolution. At the same time, sex is also constitutive of evolutionary narratives at a more fundamental level. In evolutionary narratives, sex functions both as the narrative impetus that engenders the story in the first place and as the structural engine that keeps it going. In *Ever Since Adam and Eve*, Malcolm Potts and Roger Short, for example, portray sex as

> the mainspring of our existence and the very centre of our being. It is our unseen guide along all the walks of our lives. It is the triumph of life over death, and offers our genes the promise of immortality. Sex controls our relations inside and outside our families. It accounts for many of our murders, all of our rapes, and possibly most of our wars. Our bodies and

our behaviour are the product of Darwinian evolution driven by a ruthless competition to reproduce.

(Potts and Short 1999: 1)

If all our acts, desires, and preferences are indeed end-products of "Darwinian evolution driven by a ruthless competition to reproduce," as Potts and Short insist, then sex—understood as reproductive coupling—functions as the fundamental narrative event that keeps the evolutionary narrative moving.

This narrative logic relies on adaptationism, the assumption that evolution is driven by the emergence of new beneficial traits (adaptations). This is the theoretical model that organizes sociobiology and evolutionary psychology. It is also implicit in a number of other evolutionary narratives, some of which openly critique evolutionary psychology, as we will see in the following chapters. Perhaps the most peculiar feature of adaptationist thinking is the curiously circular narrative logic it produces. Potts and Short's text, for example, understands reproductive acts as productive of both new beneficial traits—adaptations—and the continuation of the evolutionary narrative. Reproduction and adaptation, that is, are implicated in one another so that adaptations lead to reproductive success while reproduction appears as the precondition of adaptation. At the heart of contemporary evolutionary narratives, then, is the endlessly repeated sequence of reproductively successful sex acts.

Many popular texts on evolution openly acknowledge the constitutive function of sex in the narrative dynamic of evolution. In *The Evolution of Desire*, evolutionary psychologist David M. Buss declares that

[t]hose in our evolutionary past who failed to mate successfully failed to become our ancestors. All of us descend from a long and unbroken line of ancestors who competed successfully for desirable mates ... We carry in us the sexual legacy of those success stories.

(Buss [1994] 2003: 5–6)

For Buss, sex functions as the endlessly repeated narrative event that turns certain organisms into ancestors. At the same time, sex is also the ultimate outcome of the evolutionary narrative, manifest in the "sexual legacy" of the prehistoric past that "we carry in us." Buss' text highlights how sex as a constitutive narrative event produces sex as thematic content. Such a dynamic generates a narrative trajectory that seems to originate in the Pleistocene—the prehistoric period when "human nature" allegedly became established—and extends, through the uninterrupted sequence of reproductive sex acts, to the present. This trajectory, however, is not really a trajectory at all. Rather, it is a self-replicating narrative loop in which the story always coils back to the originary moment, the establishment of sex as the constitutive narrative event, in order to reproduce the narrative about how that same sex came to control our twenty-first-century lives.

Modes of critique

Evolutionary discourse has been an object of feminist critique especially since the rise of sociobiology in the 1970s and its reincarnation as evolutionary psychology in the 1990s. These critiques have typically followed two lines of argument. On the one hand, critiques have identified heterocentric and anti-feminist—even misogynous—assumptions in evolutionary arguments made by sociobiologists and evolutionary psychologists, thereby aligning the two projects with reactionary politics. These critiques often focus on the representation of gender and sexuality in the popular science literature produced by, or surrounding, sociobiology and evolutionary psychology. Sophisticated critiques of this kind have been made, for example, by Roger N. Lancaster (2003; 2006), Susan McKinnon (2006), and Martha McCaughey (2008). Second, and often simultaneously, such critiques have identified false assumptions in the choice of method, the collection of data, and the analysis of results in the science itself. Critiques of this second kind typically portray sociobiology and evolutionary psychology as bad science—or even as pseudo-science. Key contributions have often come from feminist and queer scientists, such as Ruth Bleier (1984), Anne Fausto-Sterling, Patricia Gowaty and Marlene Zuk (1997), and Joan Roughgarden (2004).

My book springs from this same concern about the popularity of evolutionary explanations of gender and sexuality. At the same time, it is also underwritten by a growing feeling that current critiques of evolutionary psychology may not have captured some of its crucial characteristics. I feel that by equating sociobiology and evolutionary psychology with reactionary politics, feminist critique may risk reducing a highly complex phenomenon to a single level of explanation. If particular scientific arguments are entangled with particular cultural values, what exactly is the nature of this entanglement? What is the intertextual dynamic through which evolutionary arguments about gender and sexuality appropriate and reinforce ideological frameworks? Similarly, while methodological critique of sociobiology is crucial to contesting evolutionary psychological claims, such critique is unable to address the curious persistence of particular evolutionary arguments. Nor does it explain the intertextual dynamic through which evolutionary knowledge operates in historically specific formations of culture.

In this book, I interrogate the cultural appeal of contemporary evolutionary discourse by tracing its epistemic, discursive, and narrative commitments. I suggest that instead of thinking of popular evolutionary discourse simply in terms of liberal versus conservative values, or good science versus bad science, we need to examine its investments in such concepts as foundationality, teleology, irreversibility, reproduction, mutation, innateness, and multiplicity—all central and contested concepts in evolutionary thought. These concepts have implications for ongoing debates about epistemic authority, cultural privilege, and ontological categories. They also reshape our understanding of the past and the future, origins and destinies, progress and regress, and transformation and stability. In order to understand the narrative appeal of evolutionary claims of sexuality, we need to understand the ways in which such claims revise these apparently nonscientific issues.

This approach enables me to ask questions that other methods of analysis could not easily address. For example, if a narrative is an account of change over time, and a narrative that refuses movement seldom counts as a "good story," where exactly do evolutionary psychological narratives of immutable sexuality derive their rhetorical power? If narrative is a product of culture, are structural limitations on narrative transformation indicative of the wider parameters of culture change? The chapters of this book demonstrate that narrative analysis not only helps us to understand the cultural appeal of evolutionary narratives of sexuality, but also allows us to identify crucial omissions and contradictions in their narrative logic. Such narrative inconsistencies provide discursive sites from where feminist and queer projects of rewriting may potentially emerge.

As my aim is to identify and examine those lines of narrative descent that are constitutive of contemporary discourses of evolution, I focus on narratives that resonate with popular debates about gender, sexuality, and human nature. Although there are a number of scientists who question strict adaptationism, their alternative accounts of evolution have gained much less popular attention than those produced by evolutionary psychology. For example, Stephen Jay Gould's, Niles Eldredge's, and Richard Lewontin's contributions to both evolutionary theory and its popularization have been considerable (Eldredge and Gould 1972; Gould and Lewontin 1979; Eldredge 1985; Gould [1989] 1990; Eldredge 2004). Yet their ecologically tuned revisions of evolutionary change are practically absent from the realms of entertainment and advertising. This is why I focus largely—though not exclusively—on sociobiology and evolutionary psychology.

Narrative and knowledge

I examine evolution as a narrative because this allows me to capture tensions between the structural, contextual, and discursive in the cultural understanding of science. Literary scholar Gillian Beer (1999: 173–95) argues that the influence of scientific theories on the cultural imagination is often subtle, present in narrative form rather than an explicitly articulated theme. Like Beer, I understand narrative as a phenomenon that underlies both scientific and popular modes of knowledge production. Narrative is a site through which scientific ideas enter cultural discourse, and cultural discourse seeps into science. It thus serves as a nodal point through which the various cultural resonances of evolutionary theory may be interrogated.

Narrative is also central to our understanding of science. In her study of the cultural aspects of genomic knowledge, cultural studies scholar Judith Roof observes that DNA and genes are often invested with "a narrative power, one of the most basic, pervasive, and least-acknowledged processes in contemporary culture" (Roof 2007: 18). For Roof, narrative is ultimately a tool that produces shape, "a way to arrange cause and effect to produce a linear, sensible scenario" (Roof 2007: 18). Such a tool makes science resonate with our understanding of the material world and our own cultural context. As a result, "science itself is transformed from a complex set of processes, systems, and stages to a matter of instant and naturalized causality" (Roof 2007: 18). Cultural studies scholar José van Dijck highlights this role of narrative as a technology of knowledge. For van Dijck, the cultural significance of narrative arises from the fact that "it serves as a mode of cognition without it being itself recognizable as an interpretive framework" (van Dijck 1998: 15).[4] Focusing on narrative in popular evolutionary discourses, then, makes it possible to reach beyond the scientific content (the set of claims) to the underlying logic that structures cultural representations of scientific knowledge.

Apart from facilitating access to the implicit and the structural, narrative analysis is also sensitive to the ways in which popular accounts of science rely heavily on narrative. As consumers of popular science—including such seemingly nonscientific genres as news reports, talk shows, or Hollywood movies—we are familiar with narratives of heroic scientists, protective lion mothers, scheming viruses, and Machiavellian genes. In his analysis of professional and popular science articles, Greg Myers suggests that popular science tends to employ what he calls a narrative of nature, a model "in which the plant or animal, not the scientific activity, is the subject, the narrative is chronological, and the syntax and vocabulary emphasize the externality of nature to scientific practices" (Myers 1990: 142). Within this narrative framework, nature appears as an anthropomorphic kingdom populated by human-like organisms, while the methodological steps through which that kingdom appeared as an object of study simply disappear. This strong pull of narrative is also present in scientific detective plots identified by Ron Curtis (1994) in popular science magazines, and examined by Priscilla Wald (2008) in the "outbreak narrative," the narrative of the emergence, spread, and containment of infectious diseases. The detective plot represents the scientists as unrelenting detectives who investigate perplexing mysteries, such as the identity of a rare virus. Both Curtis and Wald demonstrate that detective plots are highly invested in ideology, as the narrative dynamic of adventure and discovery, or crisis and resolution tends to render certain views and viewpoints epistemically privileged.

While the prominence of narrative in popular science has been noted widely, there is no consensus as to what is meant by narrative. Many studies on scientific narratives understand narrative in its common-sense meaning of a "story" or an "account." Even when examined as a particular causal arrangement of observations and experiences, narrative still tends to be located at the level of content. This is the case, for example, with Donna Haraway's *Primate Visions*, which explores science as "a story-telling practice," suggesting that "[b]iology is inherently historical, and its form of discourse is inherently narrative" (Haraway 1989: 4). Similarly, Roger Lancaster often uses the word *story* in an untheorized way, as when he views evolutionary psychological accounts of gender and sexuality as "modern origins stories" (Lancaster 2003: 38). While Haraway insightfully captures the processes through which scientific explanations appear as fact, and Lancaster shows how the scientific under-standing of sexuality arises from and feeds back into cultural ideas of identity and agency, neither of them attempts to provide a more structural understanding of narrative in science. The persuasiveness of scientific narra-tive thus becomes located in the interface between scientific practice and cultural expectations rather than in the narrative dynamic itself. Unlike Haraway and Lancaster, I view narrative as a primarily structural phenom-enon, as an underlying logic that organizes accounts of science and its dis-coveries. From this point of view, any given arrangement of narrative content emerges from and participates in a structural dynamic that always exceeds that arrangement. Yet narrative-as-structure cannot be separated from its historical and cultural contexts, as structure is both persistent and open to change.

My approach is based on the premise that narrative is always intertwined with cultural discourses. This embeddedness is not a matter of ideological coloring that can be located in the narrative content or its linguistic expression and thus removed from the text. Instead, ideology permeates the very principles of narrative organization, with the result that narrative participates in the processes that produce knowledge and identities. Narrative is not, however, opposed to fact in the sense that it would produce mere pseudo-science or folk wisdom. Haraway makes this clear in her intriguing discussion of the words *fact* and *fiction*:

> Facts are opposed to opinion, to prejudice, but not to fiction. Both fiction and fact are rooted in an epistemology that appeals to experience. How-ever, there is an important difference; the word *fiction* is an active form, referring to a present act of fashioning, while *fact* is a descendant of a past participle, a word form which masks the generative deed or perfor-mance. A fact seems done, unchangeable, fit only to be recorded; fiction seems always inventive, open to other possibilities, other fashionings of life.
>
> (Haraway 1989: 4)

The main difference between fact and fiction is one of temporality. That is, it is a difference produced through retrospective distance. While fiction does not always generate facts—very often it does not—fact is always a product of processes of fiction. It is precisely because fiction is not the counterpart of fact but its constitutive principle that narrative permeates our cultural practices and interpretative models and is so thoroughly intertwined with cultural dynamics of privilege and authority. This idea of narrative (or Haraway's fic-tion) as a process that naturalizes assumptions provides a starting point for my analysis of evolutionary narratives.

Gendering narrative

Feminist narrative scholars have demonstrated that narrative is intrinsically intertwined with gender ideologies. Cultural studies scholar Bernice L. Hausman (2000) maintains that narrative is one of the primary mechanisms through which assumptions about biological gender differences are produced as scientific fact in popular discourses. If narrative naturalizes ideas and experiences, then narratives of gender turn gendered assumptions into gendered facts. Accordingly, Hausman argues, narrative analysis may help denaturalize commonsense assumptions about gender by capturing gender as it is being constructed. As a result, gender emerges as "a dynamic category of subjectivity, rather than a static referent of known contents" (Hausman 2000: 117), and gendered categories become understood "as ideas rather than as facts" (Hausman 2000: 118).

This kind of narrative analysis does not suggest that gender is "just" narrative. Gendered embodiment is always more than flesh, blood, and molecules, as all accounts and experiences of the gendered body take shape within the conceptual frameworks of culture. Yet narrative is inseparable from the material body, as gendered experience cannot be reduced to either language or biology. Narrative, then, is a cultural mechanism through which bodies become understood as seats of gendered identities. It is a structural dynamic through which gendered bodies appear as objects of knowledge. It is a technology of knowledge that produces gender as a binary ontological category and the gendered body as the precondition of all experience.

Narrative is also gendered on a structural level. Feminist psychoanalytical scholars in particular have long drawn parallels between narrative form and pleasure, suggesting that the appeal of narrative arises from its correspondence with heterosexual male desire and the masculine logic of tumescence, climax, and discharge (de Lauretis 1984; Winnett 1990). Female desire, this suggests, can find expression only through unconventional narrative structures, or through other modes of language such as experimental poetry. Other scholars have questioned the universalistic assumptions that underlie this equation of male desire and narrative (Clayton 1989; Homans 1994; Friedman 1998; Felski 2003; Page 2007). For these critics, too, gender and narrative are intertwined at the level of narrative structure, so that certain narrative patterns tend to reinforce certain ideological positions. However, these narrative patterns are culturally specific and historically changing, so that no shape of plot has a fixed meaning. This also means that narrative patterns can be revised and rewritten, even if such reworking is not easy, and its effects may be subtle and unstable.

Queer scholars have drawn attention to the ways in which narrative naturalizes particular models of heterosexuality. For example, Judith Roof argues that "our very understanding of narrative as a primary means to sense and satisfaction depends upon a metaphorically heterosexual dynamic within a reproductive aegis" (Roof 1996: xxii). In such a dynamic, narrative entities—characters, forces, opinions, ideologies, ways of life—come together in order to produce more narrative. This promise of narrative futurity in turn reinforces the primacy of reproductive heterosexuality that justified the narrative dynamic in the first place. For Roof, this connection between narrative structure and reproductive heterosexuality is first and foremost metaphorical, which is why it is so culturally persuasive. Queer literary scholar Dennis W. Allen (1995) takes a more optimistic stance, as he suggests that the equation of heterosexuality with narrative and homosexuality with the impossibility of narrative is far from stable. For Allen, "narrative is less a reflection of sexuality than a process of constructing it" and the narrative about the naturalness of heterosexuality "is performative rather than mimetic" (Allen 1995: 627). While narrative is more likely to resonate with some stories (those of heterosexuality) than others (those of homosexuality), narrative is not aligned with a specific ideology. Thus we are able to question "the ubiquity and inevitability of both heterosexuality and its stories" by challenging the naturalized link between heterosexuality and narrative (Allen 1995: 627; see also Farwell 1996; Lanser 1996, 2009).

This book investigates evolutionary narrative both as a persuasive pattern organized by a set of premises about gender and sexuality and as a productive dynamic that generates facts about gender and sexuality. The book seeks to capture the tension between the persistence of particular narrative structures and the variations of narrative they nevertheless engender. Through this focus on the charged interface between narrative stability and mutability, the book traces evolutionary narratives' changing commitments to ideologies of gender and sexuality.

Across genres

The chapters that follow trace evolutionary narratives through a range of texts representing an array of genres. These genres include scientific articles, popular science books, popular science magazines, news reports, popular fiction as well as literary works. This focus situates my investigation within the debates about the popularization of science. In public discourse, popularization is commonly assumed to be a matter of straightforward movement from moderate and rational professional science to its unruly and unpredictable simplification. While it is widely acknowledged that scientific communities need popularization to communicate their results to the nonscientific public—and to the people and institutions that make future funding decisions—popularization is viewed as at best producing a watered down version of the original study, and at worst generating misinterpretations.[5]

Scholars studying popularization have argued that professional and popular science form a continuum rather than a strict dichotomy. First, the idea of a purely professional research paper emerged only at the end of the nineteenth century (Montgomery 1996: 25–31; Myers 2003: 268; Paul 2004: 34–7). Before that, scientific texts tended to address an audience that included professional academics, amateur scientists, as well as a wider "gentleman" audience. Darwin's work is a case in point: while participating in the heated scientific debates of the day, he addressed a general educated audience and often employed non-specialist language that was rich in literary influence. This historical perspective leads linguistic scholar Greg Myers to conclude that although the increased professionalism and specialization among scientists have been crucial factors in the rapid growth of science during the past hundred years, "they were not inevitable or intrinsic to the subject matter of science" (Myers 2003: 268). Viewed within this historical context, differences between popular and professional genres of science emerge as being of degree rather than kind.

The different genres of science writing also demonstrate considerable internal variation, so that a given genre may include a range of overlapping strategies to suit various debates and cultural contexts. This suggests that the audiences addressed by popular science texts are far from homogeneous. What is often referred to as a "lay audience" may turn out on closer inspection to also include non-professional readers with a high competence in a particular field (such as computer science enthusiasts) and professionals in intermediate positions (doctors and legislators, for example). The audience also often includes practicing scientists working in related fields: a primatologist may read a popular book by a molecular geneticist, a cognitive linguist a popular article by an evolutionary ecologist. Rhetorical scholar Danette Paul points out that this phenomenon is a result of the specialization of science during the past century, which has made it difficult for scientists to follow the advances and debates in other fields of science (Paul 2004: 36). But genres with multiple audiences are not absent at the other end of the professional–popular continuum either. For example, many scientific journals publish reviews of research on a particular topic, an in-between genre read by professional scientists as well as policy makers and other professionals outside academia (see especially Hilgartner 1990). The line between professional and popular science, then, is far from clear-cut.

Furthermore, cultural hopes, fears, and beliefs about science do not arise solely, or even primarily, from popularizations that aim to educate or persuade. Cultural artifacts like films, television shows, visual art, and fiction often appropriate science-related headlines to their own ends, highlighting expectations and anxieties rather than factual claims. While suggesting scenarios perhaps never intended by scientists, such reworking of scientific knowledge affects the ways in which the role of science is understood by individuals, institutions, or demographic groups. As Priscilla Wald observes, fictional genres differ not only from professional science but also from the news media in that they may "dramatize the implications and extend the metaphors into fully articulated scenarios. Even when the science seems incidental—new seasoning for the familiar recipes—the enactments display and reinforce the logic of the terms through which the issues are defined and debated" (Wald 2005: 212). An analysis of fictional genres may thus help capture implicit tendencies and contradictions in science itself. By examining evolutionary narratives across a range of genres, I hope to shed light on the web of cultural assumptions and narrative expectations that underlies popular discourses of evolution.

The structure of the book

The chapters that follow trace the emergence and consequences of the discursive shift that turned Darwin's understanding of "utter licentiousness" as a sign of a lower evolutionary stage into genetically coded profligacy characteristic of the whole human species. The chapters demonstrate that sociobiological and evolutionary psychological narratives appropriate and extend the role assigned to reproduction in Darwin's writing. This engenders a reproductive narrative logic that imagines the desire to reproduce as the driving force of evolution and posits the reproductive sex act as the endlessly repeated narrative event that keeps the story going. My investigation suggests that the popular appeal of contemporary evolutionary explanations of gender and sexuality is not simply symptomatic of general antifeminist or reactionary tendencies in culture. Rather, it arises from the reproductive structural dynamic that underlies evolutionary narratives. That dynamic, however, is inherently inter-twined with historically specific cultural discourses and thus with ideology.

The book also demonstrates that this reproductive narrative logic is structurally vulnerable. Since the continuation of the evolutionary narrative relies on successful reproduction, the possibility of reproductive failure poses a constant risk for the evolutionary narrative. Furthermore, the very distinction between reproduction and nonreproduction is fundamentally unstable. Despite the politics of immutability that characterizes contemporary evolutionary discourse, then, neither evolutionary narratives nor the genders and sexualities they produce are ever fully fixed. I argue that this intrinsic structural weakness in the evolutionary narrative logic explains some of the cultural anxieties that evolutionary ideas raise. It also produces a tension between the politics of stability that evolutionary psychology advocates and the structural striving for variation and change that underlies its central narratives.

The chapters of this book explore this intricate narrative dynamic, its persuasive power, and its inherent contradictions. Chapter 1, "Foundational Trajectories in Darwin and Sociobiology," sets out to reevaluate and clarify the connection between Darwin and sociobiology in terms of their narrative commitments. Through my reading of Darwin's *The Origin of Species* (1859) and *The Descent of Man* (1871), and E. O. Wilson's *On Human Nature* (1978), I trace the ways in which both Darwin and Wilson imagine evolution as the ultimate metaphysical narrative, connecting assumptions of change and stability to the politically resonant ideas of foundationality and teleology. Furthermore, by examining Darwin's and Wilson's appropriation of religious, mythic, and nationally coded discourses, I outline the specific narrative strategies through which the two theorists produce their evolutionary narratives as epistemically privileged accounts of origins and destinies. These affinities and discontinuities between the two theorists shed light on the different politics of gender and sexuality that emerge in evolutionary narratives today.

Chapter 2, "Narrative Variation and the Changing Meanings of Movement," shifts the discussion of foundationality, teleology, and the preconditions of change to the present. The chapter focuses on two mutually contradictory appropriations of Wilson's evolutionary epic: environmentalist "Epic of Evolution" texts that evoke evolution as a source of ethical guidance and spiritual fulfillment, and openly atheist, "philosophical naturalist" texts that posit matter as the ultimate level of explanation. Despite the stark difference in their tone, both traditions of evolutionary narration appropriate the ambiguous foundational potential in Darwin's writing, unsettling rather than resolving the ambivalent relationship between evolutionary theory and assumptions of chance and mutability. The chapter connects these variations of evolutionary narrative to the long-standing controversy over evolution and creationism in the US. The second part of the chapter examines two novels, John Darnton's *Neanderthal* (1997) and Will Self's *Great Apes* (1997), in order to understand the cultural reception of different versions of evolutionary foundationality. These novels show how claims of foundationality and epistemic authority take shape through nationalistically and religiously coded discourses. They also demonstrate that ideologies of movement are paradoxically often articulated through assumptions of immutable sexuality.

Chapter 3, "The Gendered Politics of Genetic Discourse," investigates how the eon-spanning foundational narrative explored in the previous chapters is reimagined as a molecular evolutionary narrative in the second half of the twentieth century. Tracing this molecular epic back to the discovery of the DNA structure, the publication of Richard Dawkins' *The Selfish Gene*, and advances in imaging technologies, the chapter analyzes the consequences of this molecular-level narrative on assumptions about the mutability of sexual desire and gendered behavior. Through my reading of such texts as Bryan Sykes' popular science book *Adam's Curse*, Natalie Angier's popular science book *Woman*, and Simon Mawer's novel *Mendel's Dwarf*, I suggest that the molecular imaginary provides a narrative dynamic that resolves structural contradictions that haunt the Darwinian evolutionary narrative. Most importantly, they provide a coherent narrative agent: the gene. This molecular evolutionary narrative turns molecular entities into anthropomorphic and gendered agents, thereby naturalizing gender ideologies. At the same time, it also points to an inherent tension between the way in which the reproductive narrative logic embraces mutation and adaptability, and the assumption of immutable gender and sexuality that this logic nevertheless engenders.

Chapter 4, "The Narrative Attraction of Adulterous Desires," takes a closer look at the reproductive narrative logic by examining the ultimate expression of reproductive ambition: sexual infidelity. The chapter focuses on the underlying structure of the evolutionary infidelity narrative, the narrative of the emergence, persistence, and primacy of promiscuous behavior within an evolutionary framework. Through popular science texts such as Matt Ridley's *The Red Queen* and Olivia Judson's *Dr. Tatiana's Sex Advise to All Creation*, the chapter identifies a feminist and an anti-feminist version of this narrative, which both imagine evolution as a chain of reproductively motivated acts. The chapter then turns to three novels—David Lodge's *Thinks … ,* Alison

Anderson's *Darwin's Wink*, and Jenny Davidson's *Heredity*—which all point to a parallelism between the structures of the infidelity and romantic narrative. Such parallelism suggests that the evolutionary infidelity narrative relies on a constant repression of the possibility of reproductive failure, and that the narrative structure of romance serves as a means of negotiating this intrinsic instability.

Chapter 5, "Reproductive Failure and Narrative Continuity," examines the polar opposite of reproductively motivated licentiousness: nonreproductive sexuality. If infidelity is understood as the ultimate maximization of one's reproductive fitness, then sexual practices and desires that do not result in reproduction run against this narrative logic. The chapter explores the many forms of exclusion that the reproductive imperative as a narrative engine generates through four case studies: scientific explanations of female orgasm and male homosexuality, the persistence of nonreproduction in Jeffrey Eugenides' novel *Middlesex*, and the prominence of identity discourse in the media coverage of scientific studies of sexuality. The chapter identifies an event-based narrative logic in adaptationist evolutionary narratives that first produces nonreproductive sexual acts, pleasures, and desires as a logical impossibility and then proceeds to integrate them. This narrative logic is, however, limited by its own privileging of events, which leaves desires, pleasures, and identities outside its explanatory scope.

The concluding chapter addresses the charged interface between narrative form and cultural context. This interface between structure and context is where the constraints and potential of evolutionary narration are negotiated. As a contested territory, it is also a site from where feminist and queer projects of rewriting may arise.

Notes

1 For critiques of essentialist assumptions in the popular genetic discourse surrounding the Human Genome Project, see, for instance, Nelkin and Lindee [1996] 2004, van Dijck 1998, van Dijck 2000, Marchessault 2000, Shea 2001 and Roof 2007. For critical discussions of sociobiology and evolutionary psychology, see Haraway 1989, Haraway 1991, Sperling 1991, Dusek 1999, Zuk 2002, Lancaster 2003, and Lancaster 2006.

2 The phrase "the profligacy of women" is repeated four times on pages 214–18.

3 *Walking with Cavemen* was part of a series of television documentaries, which also included *Walking with Dinosaurs* (1999), *Walking with Beasts* (2001), and *Walking with Monsters* (2005). *Walking with Cavemen* was subsequently shown on the Discovery Channel in the US. The American version turned the original four 30-minute episodes into two one-hour episodes and replaced the original narrator, professor Robert Winston, with the actor Alec Baldwin.

4 See also Clare Hemmings' (2011) insightful discussion of the role of narrative in feminist scholars' accounts of the history of feminist theory. Hemmings demonstrates that the narrative trajectories that feminist histories rely on operate as highly persuasive epistemic claims.

5 For a comprehensive survey of debates about scientific popularization and the need for "public understanding of science," see Gregory and Miller (1998).

1 Foundational trajectories in Darwin and sociobiology

Debates about the scientific validity of sociobiology and evolutionary psychology often turn to Charles Darwin. In popular discourses of science in particular, sociobiology and evolutionary psychology are typically portrayed through such phrases as "as Darwin already recognized," "as Darwin showed," or, alternatively, "unlike Darwin." Not surprisingly, proponents of sociobiology and evolutionary psychology tend to equate sociobiology with the cultural authority associated with Darwin. For example, evolutionary psychologist Robert Wright insists on calling sociobiology and evolutionary psychology "the new, improved Darwinian theory" (Wright [1994] 2004: 4) and their practitioners "new Darwinian social scientists" (Wright [1994] 2004: 5). Similarly, evolutionary psychologist David Buss states that "[t]he breakthrough in applying [Darwin's] sexual selection to humans came in the late 1970s and 1980s, in the form of theoretical advances initiated by my colleagues and me in the fields of psychology and anthropology" (Buss [1994] 2003: 3). This assumed connection between Darwin and sociobiology also provides a major marketing strategy, as suggested by the title of Michael Gilbert's *The Disposable Male: Sex, Love, and Money: Your World through Darwin's Eyes* (2006), or the quote from Tom Wolfe on the cover of the 2001 Abacus edition of Edward O. Wilson's *Consilience*: "There's a new Darwin. His name is Edward O. Wilson" (Wilson [1998] 2001). In all these instances, sociobiology and evolutionary psychology appear as inevitable extensions of Darwin's theory.

Critics of sociobiology and evolutionary psychology, by contrast, have dismissed or emphasized the Darwin–sociobiology connection depending on their view of Darwin, respectively, as a progressive or a reactionary. For example, sociobiology's unrelenting critic Niles Eldredge maintains that "beginning in the 1960s, evolutionary biologists began to revamp old-style Darwinian evolutionary notions—especially 'natural selection'—expressly in terms of genes. This is where the excesses began" (Eldredge 2004: 20). Eldredge insists that "Darwin's original description of natural selection—as the effect that success in an organism's economic life has on its reproductive life—remains far and away the best way of thinking about this, the central evolutionary process" (Eldredge 2004: 26). This insistence on a theoretical disconnection between Darwin and sociobiology is also evident in such book titles as *Alas, Poor*

Darwin: Arguments against Evolutionary Psychology (2000), edited by Hilary Rose and Steven Rose, or *Getting Darwin Wrong: Why Evolutionary Psychology Won't Work* (2010) by Brendan Wallace. At the same time, there are many critics who consider both Darwin's and sociobiologists' sexual politics as conservative and thus fundamentally alike. For queer scholar Roger N. Lancaster, for instance, "Darwin's 'general law' of sexual selection" is characterized by a "fetishistic naturalization of heterosexuality," which sociobiologists and evolutionary psychologists "lay hold of and distill" (Lancaster 2003: 107).

This chapter sets out to clarify the relationship between Darwin and sociobiology by examining sociobiology's discursive, narrative, and conceptual commitments to Darwin's textual politics. In particular, the chapter identifies affinities and disruptions between Darwin's and sociobiology's reworking of ideas of transformation and stasis. In doing so, the chapter touches upon a range of seemingly nonsexual and non-gendered issues such as progress, teleology, history, origins, future, destiny, uncertainty, foundationality, religion, mythology, nationalism, imperialism, and epistemic authority. I suggest that these concepts are at the heart of ongoing cultural debates about evolution—including those of gender and sexuality. By exploring the historical intersections of these seemingly nonrelated issues, we may better understand the cultural preconditions of evolutionary models of gender and sexuality.

The chapter focuses on Charles Darwin's and pioneering sociobiologist Edward O. Wilson's textual politics. This framing leaves out a number of sociobiological texts by other writers, as well as a number of texts written in the hundred years between the publication of Darwin's *The Descent of Man* and Wilson's *Sociobiology*. However, this focus allows me to pay detailed attention to the imaginative potential of the Darwinian narrative as a cultural narrative. By examining continuities and contingencies between Darwin and Wilson, I locate a textual site where the uneasy relationship between narrative structure and cultural context is negotiated. Although I read Darwin's texts as fundamentally ambiguous and thus resonant with postmodern feminist politics, my intention is not to glorify Darwin. Rather, I wish to show the extent to which Darwin's writing is multivalent and thus the extent to which both sociobiological and contemporary evolutionary narratives rely on omissions, reversals, and shifts of emphasis. Furthermore, I do not evaluate the empirical accuracy of Darwin's or sociobiologists claims about evolution. Since a text's correspondence to material reality does not determine the influence it has on wider cultural discourses, such a project would be beside the point. Instead, I focus on discursive travels between texts and contexts. My examination of these textual moves will be guided by the idea of evolution as a foundational narrative.

Foundational narratives

The one-hundred-fiftieth anniversary of *The Origin of Species* in 2009 generated wide media coverage of Darwin's work and its contribution to today's science.

Tributes included new editions and translations of Darwin's texts, documentaries on Darwin's life, and special magazine issues dedicated to Darwin's work. The *National Geographic*, for example, published two articles on Darwin's legacy in February 2009. One of the articles, "Modern Darwins," outlined how modern science had proved correct Darwin's hypotheses (Ridley 2009). In doing so, the text inadvertently articulated many of the tensions surrounding Darwin's work in contemporary culture. Referring to the advances in molecular biology and genetics in the second half of the twentieth century, the author, evolutionary biologist and science writer Matt Ridley, writes:

> The vindication came not from fossils, or from specimens of living creatures, or from dissection of their organs. It came from a book ... To understand the story of evolution—both its narrative and its mechanism—modern Darwins don't have to guess. They consult genetic scripture.
>
> (Ridley 2009: 58)

As audiences versed in popular scientific metaphor immediately recognize, the book Ridley refers to is the genetic code written in the DNA alphabet and carried by every organism. This DNA book is not, however, just any book. It is the new scripture, the foundational account of the world that, Ridley's biblical imagery implies, challenges the authority of the Judeo-Christian creation story. The way in which Ridley invokes both "narrative" and "mechanism" further suggests that these two are intimately connected: narrative emerges from the mechanistic processes of evolution, while both narrative and mechanism unfold as a result of scientific advance.

Ridley's portrayal of evolution as the ultimate epic written in the genome did not introduce anything new to the public debates about the significance of evolutionary theory. The idea of a genetic "text" has been around since the discovery of the structure of the DNA molecule in 1953.[1] The debates about evolution and religion, in turn, date all the way back to the pre-Darwinian accounts of evolution in the late eighteenth and early nineteenth century. When Darwin published his theory of natural selection in 1859, he was faced with a long tradition of opposition to any challenge to God's direct involvement in the creation of species.[2] Darwin's critics immediately recognized the threat implicit in the idea of natural selection, which had no space for supernatural interference in the emergence and extinction of species. What raised the stakes even higher was the fact that Darwin—unlike Alfred Wallace Russell, the other theorist of natural selection—placed humanity within the mechanistic laws of nature. The way in which Darwin emphasized natural selection and the absence of direct supernatural engagement made it difficult to see humankind as an exceptional divine creation. It was precisely this dual threat to the authority of religion and the position of humanity in the Great Chain of Being, the hierarchical "ladder" of nature, that made evolution, as Daniel Dennett puts it (1996), a "dangerous idea" with fundamental implications for questions of origins, purpose, and destiny. This contested role of evolution as a foundational narrative continues to underlie today's debates about evolution, creation, materiality, and spirituality.

Cultural anxieties about the epistemic reach of evolutionary theory have not been just about religion. While sociobiologists' violent and pessimistic view of human nature has raised loud protests within academia since the 1970s,[3] these protests have been underwritten by a more fundamental concern about the association of evolution with scientific advance and epistemic authority. For example, late paleontologist Stephen Jay Gould cautions against evolutionary models that imply a narrative of progress, as they serve to "nurture our hopes for a universe of intrinsic meaning defined in our terms" (Gould [1989] 1990: 43). More sarcastic in tone, Roger Lancaster asserts that contemporary evolutionary discourse "joins what is, in American society, a perpetually vexed and tormented problem—the question of origins—with evolutionary fables that are scarcely distinguishable from creationist myths" (Lancaster 2006: 102). Both Gould and Lancaster question the assumption that evolution provides an epistemically privileged explanation of the telos of modern humanity.

My exploration begins with the premise that this potential for foundationality makes evolutionary narratives sites where the cultural authority of science can be invoked, contested, and reproduced. Turning to Darwin and Wilson, I ask where this foundational potential arises from and to what purposes it can be appropriated. Through my reading of Darwin's *The Origin of Species* and *The Descent of Man*, and Wilson's *On Human Nature*, I explore the narrative dynamic through which evolutionary narratives produce a sense of foundationality, and the ways in which that foundationality is intertwined with the rhetorical production of scientific fact. I also identify textual fractures and contradictions, which suggest that evolutionary narratives are unable to fully accommodate cultural anxieties raised by the idea of evolutionary foundationality.

Darwin's ambiguous origins

Charles Darwin's *On The Origin of Species by Means of Natural Selection, or the Preservation of Favoured Races in the Struggle for Life* was first published in 1859. The book develops the idea of natural selection as the primary mechanism of evolutionary change by bringing together Thomas Malthus' idea of scarcity as constraining growth and engendering competition, and Charles Lyell's theory of the gradualness of geological change. Through an abundance of examples of variation among wild and domesticated species, Darwin then theorizes the processes through which certain traits become prominent or disappear in populations. In 1871, Darwin published *The Descent of Man, and Selection in Relation to Sex*, which ventured into the political minefield that was the question of humanity's place in nature. Invoking again an astonishing range of examples from the natural world, the book traces similarities between humans and animals in behavior and physiognomy as well as explains the workings of sexual selection in nature. Whereas *The Origin* is primarily concerned with the questions of origins, change, and the interconnectedness of species, *The Descent* interrogates the trajectory of human evolution.

In a nutshell, Darwin's theory of descent understands species as having emerged from the accumulated variation within earlier species. Instead of using the word *evolution*, which, at the time, was usually associated with the development of an organism from a rudimentary form (such as an egg), Darwin referred to this process of change as *descent with modification*.[4] The chief mechanism through which this change takes place is *natural selection*—a term coined by Darwin—which passes on to subsequent generations those traits that are useful for survival in a specific environment. Organisms that best adapt to environmental conditions and changes in those conditions— scarcity of food, changes in climate or population density, introduction of new species or disease in the area—are the ones most likely to reproduce successfully. If the trait responsible for the organism's success is hereditary, then the offspring of that organism are also more likely to reproduce than the offspring of organisms that lack the trait—provided, again, that there are no dramatic changes in the environment that might disfavor the trait. This leads to increased differentiation both among and within species, which may eventually lead to further events of speciation. In this framework, the gradual accumulation of difference engenders new species while extinction is understood as the failure of species to respond to the changing environment.

The process of natural selection is complicated by *sexual selection*, the preference of one sex—usually the female—for particular traits in the other sex. The preferred traits are not necessarily related to survival, and indeed may seem to jeopardize survival by attracting predators or requiring an increased supply of calories—the peacock's tail is a famous example. Yet the very fact that some males are able to maintain such traits suggests health, strength, and reproductive potential. As a result, the females demonstrating the preference for such a trait in males tend to have more and healthier off- spring than other females. Since the offspring inherit their parents' traits, both the preferred trait and the preference itself may become common in the population. Sexual selection, then, may direct evolution in paths seemingly opposed to the bare logic of survival.

One of the most intense debates in Darwin scholarship has focused on Darwin's view of teleology. On the one hand, many scholars have emphasized Darwin's belief in the idea of progress characteristic of nineteenth-century liberalism. For example, philosopher of science James G. Lennox finds "a selection-based teleology" (Lennox 1993: 417) and philosopher of science Julio Muñoz-Rubio locates assumptions of "a purpose and design that respect the laws of competition" (Muñoz-Rubio 2003: 311) in Darwin's writing. Similarly, historian of science Robert M. Young maintains that "Darwin's theory was no less bound by the principle of utility than was that of [William] Paley," the famous author of *Natural Theology* (Young 1985: 97).[5] On the other hand, many scholars have drawn attention to the key role of unpredictability in *The Origin*. Feminist philosopher Elizabeth Grosz (2005), for example, identifies similarities between natural selection and the Deleuzian logic of becoming, and literary scholar Colin Milburn finds "family resemblances" (Milburn 2003: 604) between Darwin's and Derrida's understanding of origins and the logic of change. These differences in scholarly responses suggest that Darwin's view on teleology was highly ambiguous. In her seminal work *Darwin's Plots*, Gillian Beer argues that *The Origin* borrows elements from a range of cultural and literary discourses, thereby engendering associative richness that was "capable of being extended or reclaimed into a number of conflicting systems" (Beer [1983] 2000: 3). As Beer suggests, this ambiguity has contributed significantly to the heterogeneity of twentieth-century evolutionary narratives.

These debates about teleology (the assumption of predetermined trajectories) and foundationality (the assumption of fundamental consequences) are connected to the role of evolutionary origins in Darwin's writing. Darwin's discussion of descent tends to evade the question of origins, which he equates with "the dim obscurity of the past" (Darwin [1879] 2004: 679). Instead of origins, Darwin emphasizes the mechanism of evolutionary change—in *The Origin*, natural selection; in *The Descent*, sexual selection. While this emphasis points to Darwin's need to distance himself from the politically explosive debates about creation and theology, it is also indicative of the ambiguity of origins in Darwin's narrative of descent with modification. This ambivalence arises, to an extent, from the scarcity of paleontological evidence, which made a detailed discussion of the emergence of species difficult. More importantly, however, Darwin's emphasis on variation as the raw material of incipient and new species rendered the idea of a distinct point of origin questionable. If variation is the precondition of change, then that variation must have emerged from earlier variation, which must have emerged from still earlier variation *ad infinitum*.

There is a fundamental tension between heterogeneity and homogeneity in Darwin's representation of evolutionary origins. While Darwin locates origins in the singularity of "one primordial form," this singularity itself is mythic rather than factual (Darwin [1859] 1985: 455). In other words, it is singularity of a set of formal contours and behavioral patterns rather than singularity of detail. As Elizabeth Grosz insightfully puts it, origins for Darwin are "always already implicated in multiplicity or difference" (Grosz 2004: 26). This ambiguity is not merely an accident, as Darwin seems to have been aware of the consequences of this emphasis on variation. In *The Descent*, for example, he cautions against conjuring a single original couple from which humans have descended:

> It must not be supposed that the divergence of each race from the other races, and of all from a common stock, can be traced back to any one pair of progenitors ... The process would have been like that followed by

man, when he does not intentionally select particular individuals, but breeds from all the superior individuals, and neglects the inferior.

(Darwin [1879] 2004: 678)

The original singularity from which life emerged is marked by inherent multiplicity. At the same time, this point of origins cannot be fixed at any particular moment, as the chain of backward reference never reaches the instance of unpreceded wholeness. If such a moment exists, it is a fundamentally timeless and ahistorical site of beginning—which is precisely what later theories of the Big Bang have suggested.

This assumption of variance as always already there also informs the history of modern species in Darwin's writing. In both *The Origin* and *The Descent*, Darwin repeatedly rejects the idea of a specific moment of speciation. Evolutionary biologist Robert J. O'Hara (1988) has argued that it is precisely this insistence on the gradualness of change that distinguishes Darwin from both natural theology—the idea of nature as the ultimate demonstration of God's wisdom—and pre-Darwinian accounts of evolution. These earlier approaches had imagined nature as a fixed set of taxonomic hierarchies while "the things for which explanations were sought were not *changes* but rather *states*" understood in terms of their ultimate purpose (O'Hara 1988: 147).[6] Darwin, by contrast, emphasizes the gradualness of evolutionary change, thereby challenging the idea of species as a coherent entity (Beer [1983] 2000; Abbott 2003; Milburn 2003; Grosz 2004, 2005). If species have emerged through a gradual process rather than a distinct moment of speciation, then species *is* variation. Species, that is, lacks essence, as it is always constructed through retrospection. Darwin was well aware of the implications of his own critique of species. In *The Origin*, he envisions a scientific future in which "[o]ur classifications will come to be, as far as they can be so made, genealogies" (Darwin [1859] 1985: 456). This produces a two-way movement, as the search for the origin of species extends endlessly toward the evolutionary past, while the science of evolution itself seeks a future that is (at least for now) beyond its reach.

While this rejection of a clear point of origin challenges the idea of straightforward teleology, Darwin's account of descent does not fully eschew foundationality. As Colin Milburn observes, Darwin's "revisionary logic ... becomes a mythmaking of its own, a mythopoeisis deeply entangled with the imagery of monstrosity" that stands for the variation that is the precondition of change (Milburn 2003: 609). Darwin invests this alternative epic of endlessly evolving multiplicity with a sense of humility, as existing species "become ennobled" by being "lineal descendants of some few beings which lived long before" (Darwin [1859] 1985: 458). Moreover, this descent is worth respect because of its precariousness: "If any single link in this chain had never existed, man would not have been exactly what he now is" (Darwin [1879] 2004: 193). While this portrayal serves as a defense against accusations of blasphemy and degrading of humanity, it also attests to the ambiguity in

Darwin's writing. In Darwin's account of descent, there is a constant tension between the "pedigree of prodigious length" that characterizes the human species and the continuous deferral of the point of origin that the logic of natural selection necessitates (Darwin [1879] 2004: 193).

Yet this tension hardly threatens the coherence of Darwin's evolutionary narrative. The ambiguity of unfixed evolutionary origins is made safe by the epistemic scope of the theory, which is developed paradoxically through the very act of exploring origins. This claim to epistemic privilege is reinforced through what literary critic and Darwin scholar George Levine (2011) describes as a persistent contradiction between appearance and truth in Darwin's texts. Levine observes that in Darwin's writing "what *seems*, to which Darwin in his natural role as natural historian attends with the greatest care, always disguises what *is*" (Levine 2011: 26). For Darwin, the apparent and the puzzling are merely indications of a truer, deeper pattern in the natural world, which he describes with "Miltonic ambitions" (Levine 2011: 26). Darwin's vision of natural selection and speciation, then, emerge precisely through the troubling and evasive multiplicity of origins. At the same time, the ambitiousness of Darwin's "Miltonic" explanatory scope gives his evolutionary narrative a sense of foundationality. However, this foundationality is not necessarily teleological, since teleology emphasizes a distinct point of origins from which a temporal and causal trajectory may spring.

Doubtful destinies

Like his portrayal of origins, Darwin's emphasis on the interconnectedness of all life generates an uneasy coexistence of conceptual instability and foundational temporal procession. On the one hand, the idea that all life forms are interconnected indicates a blurring of borders. Darwin and his contemporaries were able to imagine in only extremely vague terms what we today know as deoxyribonucleic acid (DNA). Yet he strongly believed that all species are formed out of the same shared raw material, so that there are no categorical or ontological differences among species. As a result, any evolutionary subject, whether mollusk, chimpanzee, or modern human, is connected through "chains of affinities" to all other evolutionary subjects (Darwin [1859] 1985: 454). This continuity across species bridges space (oceans, continents) as well as time (generations, evolutionary eons).

One of the most fundamental consequences of this interconnectedness is the removal of *Homo sapiens* from the pedestal it has claimed for itself. Having "descended from a hairy, tailed quadruped, probably arboreal in its habits" (Darwin [1879] 2004: 678), *Homo sapiens* is first and foremost a primate species, and "the difference in mind between man and the higher animals, great as it is, certainly is one of degree and not of kind" (Darwin [1879] 2004: 151). Indeed, Darwin suggests, "[i]f man had not been his own classifier, he would never have thought of founding a separate order for his own reception" (Darwin [1879] 2004: 176). Rather than a narrative about humanity, Darwin understands evolution as a narrative of the emergence of the natural world. Milburn captures this sense of conceptual displacement when he observes that Darwinian "history can be written without humanity as its point of relevance" as "*Homo sapiens* is resituated within evolving nature, a system without origin, without center, boundaries or fixity, in eternal and conjugal freeplay" (Milburn 2003: 616). Yet Darwin understands nature *as history*, as a temporally structured phenomenon that is underwritten by a sense of unceasing forward movement.

Despite the lost specialness of humanity, Darwin invests his idea of evolutionary future with a potentially mythic scope. This epic sweep becomes evident in Darwin's closing remark in *The Origin*:

> There is grandeur in this view of life, with its several powers, having been originally breathed into a few forms or into one; and that, whilst this planet has gone cycling on according to the fixed law of gravity, from so simple a beginning endless forms most beautiful and most wonderful have been, and are being, evolved.
>
> (Darwin [1859] 1985: 459–60)

In this concluding sentence, Darwin connects human and mollusk destinies to the overarching narrative of life on Earth. This narrative plays with a sense of wonder that reflects Darwin's interest in the Romantics, as well as a sense of cosmological sweep that echoes Milton's *Paradise Lost*, the book Darwin brought with him on board HMS *Beagle* on his five-year journey to South America in the 1830s (Beer [1983] 2000; Levine 2011). For Darwin, the narrative of interconnected life is inherently graceful, as suggested by his emphasis on "the beauty and infinite complexity of the coadaptations between all organic beings" (Darwin [1859] 1985: 153). The enormous range of contemporaneous species constitutes a harmonious procession toward higher complexity. This sense of shared movement makes the prosaic commonness of the "hairy, tailed quadruped" appear as subordinate to the elevating grandeur of the wider evolutionary narrative. While such a highlighting of the internal continuity of nature does not depend on a teleology of change, Darwin's emphasis on beauty, harmony, and the sense of shared direction invests evolution with considerable epic potential.

Darwin's descent with modification is accompanied by a sense of irreversibility that adds to this foundational potential. Beer observes that evolution as imagined by Darwin "has no place for *stasis*. It debars return. It does not countenance absolute replication (cloning is its contrary), pure invariant cycle, or constant equilibrium" (Beer [1983] 2000: 8). The incessancy and irreversibility of change is manifest in the fact that sexual (as opposed to asexual) reproduction always generates something novel, which, over a number of generations, increases differences among organisms and species. Furthermore, Darwin's portrayal of natural selection ascribes inherent beauty to this ongoing process of transformation. Rather than "nature red in tooth and claw," as famously depicted by Alfred, Lord Tennyson, or "the survival of the fittest," as characterized by the father of Social Darwinism, Herbert Spencer (1864), Darwin's natural selection repeatedly appears as a gentle, caring, and femininely gendered power.[7] For Darwin, natural selection preserves rather than kills, "favouring the good and rejecting the bad" while "beautifully adapting each form to the most complex relations of life" (Darwin [1859] 1985: 443). There is implicit morality in such action. Understood as driven by an unselfish concern for the greater good, natural selection appears as the counterpart of the economically motivated artificial selection of animal husbandry: "Man selects only for his own good; Nature only for that of the being which she tends" (Darwin [1859] 1985: 132). Darwin's natural selection, then, is not the opposite of morality but the very articulation of its fundamental principles.

Such references to the "good" of each species place progress at the heart of Darwin's descent with modification. Although Darwin challenges the predominant understanding of nature as static, he repeatedly invokes the popular and long-standing assumption that nature constitutes an "ascending scale," even as he imagines that scale as moving in evolutionary time (Darwin [1879] 2004: 93, 106). Like many of his predecessors, Darwin suggests that "all corporeal and mental endowments will tend to progress towards perfection" (Darwin [1859] 1985: 459).[8] Muñoz-Rubio argues that Darwin associated this constant striving toward perfection with the "complexity, abundance, control, division of functions, efficiency and dominance" of organisms, all attributes that echoed Darwin's commitment to liberalism (Muñoz-Rubio 2003: 316). Reflecting the culture within which he wrote, Darwin portrays this ascending scale of complexity as culminating in "Man, the wonder and glory of the Universe" (Darwin [1879] 2004: 193), who occupies "the very summit of the organic scale" (Darwin [1879] 2004: 689). This mythic "Man" is not just any man but, in accordance with Victorian imperialist discourse, represents the "western nations of Europe" (Darwin [1879] 2004: 167). Furthermore, the mythic "Man's" moral superiority is presumably proven by "[t]he remarkable success of the English as colonists" (Darwin [1879] 2004: 167). Such terms as "good," "progress," and "complexity" hence turn out to be deeply entangled in the geopolitics of racial, gendered, and class privilege.

As a result, the evolutionary summit occupied by this intellectually and culturally privileged "Man" also emerges as a site of moral privilege. Since natural selection is a never-ceasing process, this narrative of progress also implies a possibility of further moral elevation. Darwin explains:

> Looking to future generations, there is no cause to fear that the social instincts will grow weaker, and we may expect that virtuous habits will grow stronger, becoming perhaps fixed by inheritance. In this case the struggle between our higher and lower impulses will be less severe, and virtue will be triumphant.
>
> (Darwin [1879] 2004: 150)

Such an ethically motivated evolutionary genealogy promises a release from the hold of primitive instincts associated with lower animals and the supposedly less civilized people. In spite of the multiplicity of evolutionary origins, Darwin's evolutionary narrative not only insists on forward movement—the sense of history Darwin holds on to. It also promises moral salvation through a textual politics of progress.

While this evocation of moral elevation produces a tone of mythic significance, the sense of foundationality we encounter in Darwin arises, to a large extent, from the structural organization of his evolutionary narrative. Robert J. O'Hara (1992) argues that the representation of phylogeny (the lines of descent connecting species) in both professional and popular evolutionary narratives is often characterized by anthropocentrism that turns the profusion of lines of descent into a single line leading to modern humanity. As a result, what actually resembles a branching tree without a single endpoint becomes the old Great Chain of Being rewritten as a linear narrative. This model of representation confuses irreversibility (there is no return) with linearity (there is only one endpoint). O'Hara observes that the retrospective nature of evolutionary history contributes to this confusion, as the endpoint of the evolutionary narrative determines which adaptations may appear as significant innovations (O'Hara 1992: 153–4). In an essay published in the prestigious scientific journal *Nature*, Sean Nee addresses the same phenomenon by imagining the evolutionary history from the viewpoint of microbes. Nee humorously concludes his one-page piece: "One of the huge species, *Homo sapiens*, got remarkably self-important. But when, to his surprise, a virus wiped him out, most of life on Earth took no notice at all" (Nee 2005: 429). While Darwin celebrates variety and multitude in nature, and imagines humanity as part of historical, interspecies processes, the evolution of *Homo sapiens* nevertheless emerges as the most significant achievement of evolutionary history in his writing. The direction of evolution itself thus becomes equated with the perceived progress of the last few hundred years of Western history.

Yet Darwin's texts' commitment to such false teleology is always ambivalent. As Beer observes, there is constant tension in the Darwinian evolutionary narrative between ascent and descent (Beer [1983] 2000: 6). In a curious contrast to his vision of "a still higher destiny in the distant future" (Darwin [1879] 2004: 689), Darwin reminds us that

> we do not know whether man is descended from some small species, like the chimpanzee, or from one as powerful as the gorilla; and, therefore, we cannot say whether man has become larger and stronger, or smaller and weaker, than his ancestors.
>
> (Darwin [1879] 2004: 84)

Thus, Darwin concludes, even if evolution tends to increase complexity, "progress is no invariable rule" (Darwin [1879] 2004: 166). Furthermore, while the particular intellectual and moral characteristics of *Homo sapiens* may have seemed obvious to Darwin's contemporaries, Darwin insists that they were not necessary results of the evolution of the brain: "I do not wish to maintain that any strictly social animal, if its intellectual faculties were to become as active and as highly developed as in man, would acquire exactly the same moral sense as ours" (Darwin [1879] 2004: 122). There would have been more than one possible ending to the evolutionary narrative.

The precariousness of phylogeny itself also undermines teleology. If a "grain in the balance will determine which individual shall live and which shall die,—which variety or species shall increase in number, and which shall decrease, or finally become extinct," then the steady continuity of descent is essentially a product of retrospection (Darwin [1859] 1985: 442). The mechanism of sexual selection adds a further degree of unpredictability to the evolutionary narrative, as it, in Elizabeth Grosz's words, "deviates natural selection through the expression of the will, or desire, or pleasure of individuals" (Grosz 2004: 75). Extending and complicating the definition of a beneficial trait, sexual selection shapes the trajectories of transformation within species. In this narrative framework, the future is always insecure, extinction difficult to predict, and "of the species now living very few will transmit progeny of any kind to a far distant futurity" (Darwin [1859] 1985: 459).

The narrative of descent that emerges in *The Origin of Species* and *The Descent of Man*, then, is ultimately contradictory. On the one hand, Darwin challenges the idea of a clear point of origins and the fixity of such categories as species, and imagines the world in a state of constant flux. On the other hand, Darwin invests this narrative of constant change and ever new varieties with assumptions of beauty, dignity, and moral elevation. In Darwin's narrative, the profusion in nature is in constant tension with the distinct trajectory of human evolution. In this sense, Darwin does not actually provide a foundational narrative of human nature, but a foundational narrative of the precariousness of foundationality. Through Darwin's writing, transformation emerges as both a celebrated and feared event: transformation by natural selection indicates progress—often eagerly interpreted as human progress—but it also suggests potential extinction, to which no species, including the human species, is immune. This constant balancing between the promise of progress and the threat of extinction places survival (and hence natural selection) and reproduction (and hence sexual selection) at the heart of evolution. As mechanisms of continuity, survival and reproduction are the preconditions of a narrative trajectory and thus constitutive of narrative futurity. As we shall see in Chapters 3, 4, and 5, it is especially through the idea of reproduction that evolutionary narratives become foundational accounts of gender and sexuality.

In the century and half that followed the publication of *The Origin*, commentators on evolutionary thought have evoked diverse aspects of Darwin's account of descent. In these revisions of Darwinian thinking, the multivalence characteristic of Darwin's writing often disappears. While some texts have embraced the ethical aspects of Darwin's evolutionary vision, others have downplayed the idea of morality so important to Darwin and instead emphasized the instinctual roots of all behavior—especially sexuality and aggression. These texts' affinities to Darwin's narrative dynamic are also only partial, as they appropriate and revise assumptions of transformation, stasis, foundationality, reproduction, and sexual selection implicit in Darwin's descent with modification to their own ends. Such theoretical, conceptual and textual reworking is not necessarily a shortcoming, as it has often provided a spark for theoretical and methodological innovation. Yet these acts of rewriting show that evolutionary narratives are never innocent, never purely a matter of theory.

Foundational issues in the sociobiology controversy

In 1975, Harvard ethologist Edward O. Wilson published a book on the social behavior of species. *Sociobiology: The New Synthesis* triggered an instant and intense controversy over the possibilities and limits of evolutionary theory. This unresolved controversy also underlies the current debate about evolutionary psychology. In *Sociobiology*, Wilson offers a vision of evolutionary theory as providing an explanatory frame for social organization and behaviors such as aggression, altruism, and nurturance in a wide range of species. While the book focuses on nonhuman species—most importantly insects, Wilson's own specialty—its first and last chapter argued for the inclusion of human behavior and practices within sociobiological scrutiny. The last chapter opens with a suggestion that we should "consider man in the free spirit of natural history, as though we were zoologists from another planet completing a catalog of social species on Earth" (Wilson [1975] 1982: 547). The chapter argues strongly for the fixity of human nature, maintaining that "the moralistic rules underlying" human social structures and practices "appear not to have been altered a great deal" (Wilson [1975] 1982: 554). Accordingly, the book insists that sociobiology provides an epistemically privileged approach to human societies, and that it therefore should set the parameters for the social sciences.

Wilson's central tenets were challenged by a wide group of academics and social commentators, both within and outside the natural sciences. The most vehement critique came from the Massachusetts-based Sociobiology Study Group, which was part of the liberal pro-science organization called Science for the People. The Sociobiology Study Group's sole purpose was to investigate the premises and consequences of Wilson's sociobiology, especially in terms of its political and social implications. This it did with the prestigious voice of such well-known figures as geneticist Richard Lewontin and paleontologist Stephen Jay Gould—both Wilson's colleagues at Harvard. The critics attacked Wilson for opening way for a new racist and sexist eugenics. They thus suggested that Wilson's approach did not represent the value-neutral positivist empiricism he claimed, but a position heavily influenced by conservative ideological commitments. This critique was extended and elaborated in such works as Richard Lewontin, Steven Rose, and Leon J. Kamin's *Not in Our*

Genes: Biology, Ideology, and Human Nature (1984), Mary Midgley's *Beast and Man: The Roots of Human Nature* ([1979] 1995), and Ruth Bleier's *Science and Gender: A Critique of Biology and Its Theories on Women* (1984).

Wilson's book was also received with enthusiasm by many. The initial reviews of the book by John Pfeiffer (1975) in the *New York Times* and C. H. Waddington (1975) in the *New York Review of Books* were both positive. Indeed, a month *before* the anticipated publication of Wilson's book, the *New York Times* published a first-page article by Boyce Rensberger that celebrated *Sociobiology* as "a long-awaited definitive book" (Rensberger 1975: 52). Whereas Wilson's critics protested what they considered as disciplinary imperialism, his supporters celebrated his attempt to bring the natural and social sciences within a single field of study. This widespread enthusiasm paved the way for such sociobiological works as Richard Dawkins' *The Selfish Gene* ([1976] 1999) and *The Extended Phenotype* ([1982] 1999), or David P. Barash's *Whisperings Within: Evolution and the Origin of Human Nature* (1979).[9]

Wilson's discipline of sociobiology builds on a body of previous scholarship in zoology, anthropology, ethology, and population genetics. The term "sociobiology" was coined by zoologist John Paul Scott at a 1946 conference on "Genetics and Social Behavior." Anthropology, the academic study of humanity and human diversity, provided Wilson with a cross-cultural view of early human societies, as evident in his references to anthropologists like Richard B. Lee, Irven DeVore, June Helm, or Clifford Jolly. Wilson was also strongly influenced by ethology, the scientific study of animal behavior. Outlined by Niko Tinbergen, Konrad Lorenz, and Karl von Frisch in the 1930s, ethology approaches behaviors such as aggression or courtship practices from a cross-species perspective. Furthermore, Wilson builds on the statistical models of genetic change within populations by population geneticists such as R. A. Fisher, J. B. S. Haldane, Sewall Wright, and John Maynard Smith. In particular, Wilson relies on William D. Hamilton's statistical analysis of altruistic behavior, which suggests that altruism could be beneficial for an organism, if the altruistic organism is genetically related to the object of altruism. He also invokes Robert Trivers' studies of parental investment in offspring, rivalry among siblings, and reciprocal altruism between unrelated individuals. Both Hamilton and Trivers argue that benevolent and hostile behaviors in species are best explained from a genetic viewpoint, according to which an organism always seeks to maximize its reproductive fitness—that is, the passing of its genes to subsequent generations. These mixed roots in applied statistics and the functioning of animal communities leads Donna Haraway to state that "[s]ociobiology is a communications science, with a logic of control appropriate to the historical conditions of post-Second World War capitalism" (Haraway 1991: 58). In short, Wilson's sociobiology is characterized by a strong bias for direct causation, unconscious calculation, and the ultimate rationalization of all spheres of experience.

Wilson's contested book, then, is first and foremost a synthesizing work that brought together theories and theorists from a range of established fields.

At the same time, his all-embracing vision is the point where sociobiology touches on the incendiary question of the teleology and foundationality of evolutionary processes. Wilson's attempt to unite various disciplines studying animal and human behavior and social organization under the banner of sociobiology produces a sense of inevitable movement. This movement promises a better, more informed future through the very conjoiner of the disciplines. Crucially, Wilson's way of defining the parameters of this future as based on genetics and evolutionary biology appears less as an act of trespassing on other scholars' territories than as a logical necessity. This reinforces the idea that the sociobiological evolutionary narrative defines the very conditions of knowledge.

We shall turn now to Wilson's *On Human Nature* (1978), his follow-up to *Sociobiology*. *On Human Nature* extends the idea of foundationality implicit in *Sociobiology* to questions of religion, mythology, and teleology. Wilson's insistent way of rethinking the foundational potential of evolutionary narrative in *On Human Nature* sheds light on sociobiology's narrative investments in questions of epistemic authority and cultural appeal.

The halted evolution of human nature

In *On Human Nature* (1978), Wilson explicates his claims about the genetic basis of human behavior. Rather than toning down his argument, Wilson repeats and expands all the points about human nature that had so infuriated the liberal and leftist readers of *Sociobiology*. Moreover, he also restates his earlier claim that sociobiology will take over the study of human societies and cultures. If the humanities and social sciences are to survive, it is only by accepting the models provided by the natural sciences. Through this claim, *On Human Nature* engages in both academic politics and larger cultural debates.

As its initial preposition suggests, *On Human Nature* posits itself not as a professional text like *Sociobiology*, but as "a speculative essay about the profound consequences" of the coming together of the natural and social sciences under the sociobiological approach (Wilson 1978: x). Rhetorical scholars Susheela Abraham Varghese and Sunita Anne Abraham view *On Human Nature* as an example of what they call book-length scholarly essays, a genre that, they maintain, enables scientists to engage in "vital theory-building work" while "speaking to philosophical issues" (Varghese and Abraham 2004: 226). While Varghese and Abraham see such scholarly essays as important contributions to the public understanding of science, they do not consider these texts' rhetorical aims or ideological engagements. However, popular texts by scientists are always also implicated in questions of power and authority. As Wilson's easily accessible style and broad readership implies, *On Human Nature* seeks to secure both public and institutional support for the emerging sociobiological enterprise—and considering the 1979 Pulitzer Prize for General Non-Fiction awarded to the book, this is a task Wilson carries out quite successfully. Indeed, his rhetorical success results, to considerable extent, from the speculative and visionary aspect praised by Varghese and Abraham, as the ensuing sense of philosophical depth helps construct a position of epistemic privilege.

The foundationality of *On Human Nature* is a result of a complicated textual dynamic. To begin with, Wilson's text is underwritten by the same troubled questions of linearity, causality, teleology, and change that organize Darwin's writing. This ambiguity surfaces in the tense relationship between the narratives of human evolution and the evolution of the natural world in *On Human Nature*. On the one hand, Wilson considers human evolution as inseparable from the wider narrative of the evolution of species that sets the parameters for human descent. From the very beginning, Wilson identifies sociobiology as occupying the privileged vantage point from which alone it is possible to "consider man as though seen through the front end of a telescope" and thereby "place humankind in its proper place in a catalog of the social species on Earth" (Wilson 1978: 17). Standing for the visionary skill of sociobiology, the image of the telescope establishes a metaphorical link between the ability to see across long distances (and thus distance oneself from the object of study) and epistemic privilege, while suggesting cosmological significance through the scope the image invokes. On the other hand, such parallelism of human and non-human evolutionary narratives also implies that the intellectual and social achievements of modern humanity were no obvious endpoint of the evolution of life. Echoing Darwin's observations about similarities among primates, Wilson maintains that civilization became connected with "the anatomy of bare-skinned, bipedal mammals and the peculiar qualities of human nature" simply "by accident" (Wilson 1978: 23). The apparent linearity of the narrative of human evolution, then, is an effect of retrospection: what is now arose from what was before only through a long series of probabilities, possibilities, and pure coincidences. For Wilson, as for Darwin, there is no plan for the future inherent in the origins from which that future emerged.

Despite this fundamental ambiguity, the narrative of human descent follows a neat linear trajectory in Wilson's text. In stark contrast to the messy multitude of stories captured by Darwin's famous image of an entangled bank in *The Origin* (Darwin [1859] 1985: 459), Wilson's narrative of human evolution appears as an orderly and univocal procession through a series of important adaptations:

> [T]he earliest men or man-apes started to walk erect when they came to spend most or all of their time on the ground. Their hands were freed, the manufacture and handling of artifacts were made easier, and intelligence grew as the tool-using habit improved. With mental capacity and the tendency to use artifacts increasing through mutual reinforcement, the entire materials-based culture expanded … Cooperation during hunting was perfected and provided a new impetus for the evolution of intelligence,

which in turn permitted still more sophistication in tool using, and so on through repeated cycles of causation.

(Wilson 1978: 85)

Evolution is represented here as a chain of key innovations—bipedalism, use of hands, intelligence, social behavior—which follow the structural pattern identified by O'Hara (1992). Such a pattern turns complex and chaotic phylogeny into a linear trajectory with clearly marked climactic points (O'Hara 1992: 153–4). This adaptationist dynamic renders invisible the abundance and variation that is the precondition of evolution. It also turns a slow and unruly process of change into a sudden and causally motivated emergence of new adaptations.

The way in which Wilson emphasizes the limits of unpredictability is revealing about the logic that organizes his evolutionary narrative. Wilson considers human social phenomena as characterized by the "dual track of inheritance: cultural and biological" (Wilson 1978: 78). Even though different cultures often follow different paths, "these pathways are not infinite in number; they may not even be very numerous" (Wilson 1978: 95). In fact, he argues, "the culture of each society travels along one or the other of a set of evolutionary trajectories whose full array is constrained by the genetic rules of human nature" (Wilson 1978: 207). Such talk on "evolutionary trajectories" undermines the premise that evolution is driven in unpredictable directions. In Wilson's scenario, nature is understood as a largely fixed constant that determines culture, so that culture merely acts on and refines biological necessities, as when it "gives a particular form to the aggression and sanctifies the uniformity of its practice by all members of the tribe" (Wilson 1978: 114). At the same time, Wilson's argument for fixed human nature is produced through the rhetoric of adaptive change that posits transformation as fundamental to human nature. As the emergence of adaptations requires extensive temporal scope, adaptive change nevertheless emerges as a fact of distant past rather than a change in process. Significantly, this adaptationist logic also understands transformation as reproductively motivated. As a result, human nature appears as an ultimate product of the reproductive ambitions inherent in past evolutionary processes.

In this light, it is hardly surprising that Wilson's evolutionary narrative is underwritten by ontological assumptions about gender and sexuality. For Wilson, the "conflict of interest between the sexes" that presumably follows from "gametic dimorphism" (the difference in size of sex cells) structures gender relations at all levels of social organization (Wilson 1978: 124). Similarly, the text represents the disappearance of the estrus in humans as a significant landmark that separates human and primate evolutionary histories. At the same time, the text imagines the human disappearance of estrus as engendering further key inventions, including a fundamental change in sexual relations that established "the pair bond" and "reduced aggression among the males" (Wilson 1978: 141). What adds to the sense of foundationality produced by this narrative logic is the obscuring of the distinction between gender and sexuality. Sexuality (for example, a sexual arrangement like human pair bond) is seen as arising from gender (two conflicted yet complementary genders), which, however, is already an articulation of a particular arrangement of sexuality (a particular form of reproductive heterosexuality presumably arising from gametic dimorphism). As a result, gender and sexuality appear as articulations of each other, defined through one another in a closed circle.

There is a curious sense of narrative incongruity here, as the evolutionary process is characterized by a clear, linear trajectory, while the meta-level theoretical reasoning moves in endlessly repeated circles of causality. This contradictory movement blurs the relations of causality, as the foundational linearity of Wilson's evolutionary narrative appears as correcting the moments of methodological circularity in his text. In this sense, the book acts out in its textual politics what science studies scholar W. R. Albury (1980) identified at the time as sociobiology advocates' tendency to turn ideological critique into a matter of methodological opportunity. As Adbury demonstrates, this strategy makes methodology an issue that will be addressed in the future, but that need not bother the pioneers of sociobiology.

All in all, Wilson's emphasis on the immutability of human nature produces a contradictory representation of evolution as both historical and ahistorical. The way in which evolution appears as a series of key adaptations effects a sense of progressive movement from the past through the present into the future. At the same time, human nature is understood as a product of the prehistoric past so that implicated in the future is a constant return to the past. This tension between movement and immobility, and progress and fixity, becomes evident when Wilson proclaims that "no species, ours included, possesses a purpose beyond the imperatives created by its genetic history" (Wilson 1978: 2). While such a statement does admit the possibility of purpose, that purpose is genetic rather than metaphysical, and thus located in our past inheritance. Purpose, then, emerges paradoxically as a domain of the past. Accordingly, Wilson insists that "the urban middle class aches for a return to a simpler existence" that resembles our evolutionary roots (Wilson 1978: 93). If modern life is indeed doomed to fail to respond to "the deepest needs of human nature," as Wilson maintains, then there is no future beyond the past (Wilson 1978: 209).

A better epic

The tension between the linear progress of the evolutionary narrative and the sense that that narrative is fundamentally false produces a fracture in the logic that organizes Wilson's text. *On Human Nature* negotiates this narrative friction by evoking the idea of evolution as an epistemically privileged mythology that is based on scientific materialism. According to Wilson, such mythology's "narrative form is the epic: the evolution of the universe from the big bang of fifteen billion years ago through the origin of the elements and celestial bodies to the beginnings of life on earth" (Wilson 1978: 192). What makes this all-encompassing materialism mythology is that it "can never be definitely proved to form a cause-and-effect continuum from physics to the social sciences, from this world to all other worlds in the visible universe, and backward through time to the beginning of the universe" (Wilson 1978: 192).

Wilson portrays this evolutionary epic as "the best myth we will ever have" (Wilson 1978: 201), as it brings the overwhelming multitude of detail in the natural and human worlds under the unifying touch of a foundational narrative. However, this epic is not devoid of spiritual meaning, as its narrative seeks to satisfy the "mythopoeic drive" characteristic of *Homo sapiens* by providing an ethical and philosophical framework for human existence (Wilson 1978: 200). Echoing the sense of awe and beauty present in Darwin's writing, Wilson's text imagines the evolutionary epic as providing "the stimulus to imagination" (Wilson 1978: 205), and a view of origins that is "far more awesome than the first chapter of Genesis or the Ninevite epic of Gilgamesh" (Wilson 1978: 202). Wilson departs from Darwin, however, in maintaining that an "epic needs a hero" (Wilson 1978: 203). If we choose to read Darwin's account of descent as a foundational narrative, it is first and foremost a narrative of the natural world in which no single independent and adventurous agent plays the lead (the "hairy, tailed quadruped" hardly counts as an epic hero). Wilson's narrative, by contrast, is primarily an epic of the human condition, in which the inventive and increasingly sophisticated human mind plays the hero, whose advances the narrative follows. From these two narrative strategies arise, on the one hand, the sense of ambiguity in Darwin's treatment of teleology and foundationality, and, on the other hand, the sense of coherence and unity in Wilson's account of human descent.

Wilson's focus on the evolution of the mind allows him to position science at the final climactic point of his evolutionary narrative. Literary scholar Patricia Waugh has examined late twentieth-century evolutionary interpretations of what she calls the "metaphysics of materialism" (Waugh 2005: 242). Waugh observes that such metaphysics produces "an account of the very instrument that allows science to offer its own account" (Waugh 2005: 242). As a result, scientific inquiry "appears to break out of incompleteness and undecidability, for in reconciling mind and matter, science can claim to have arrived at that final theoretical closure which includes in its account of everything, an account of itself" (Waugh 2005: 242). Consistent with this framing, Wilson represents scientific inquiry as the culmination of evolution. For Wilson:

> Pure knowledge is the ultimate emancipator. It equalizes people and sovereign states, erodes the archaic barriers of superstition and promises to lift the trajectory of cultural evolution. But I do not believe it can change the ground rules of human behavior or alter the main course of history's predictable trajectory. Self-knowledge will reveal the elements of biological human nature from which modern social life proliferated in all

its strange forms. It will help to distinguish safe from dangerous future courses of action with greater precision.

(Wilson 1978: 96)

Science does not appear here simply as a release from the hold of evolutionary history but as "self-knowledge" that enables humankind to plan a future course of action that fully accommodates our innate needs. This science, of course, is synonymous with the emerging field of sociobiology, through which "a proper foundation can be laid for the social sciences, and the discontinuity still separating the natural sciences on the one side and the social sciences and humanities on the other might be erased" (Wilson 1978: 195). Sociobiology, then, stands as a promise that the evolutionary narrative will successfully negotiate the discrepancy between modern life and prehistoric human nature. Hence it also operates as a guarantee of narrative futurity.

This representation of sociobiology as the climax of the evolutionary narrative is involved in what Slavoj Žižek calls "the struggle for intellectual hegemony" that determines "who will occupy the universal place of the 'public intellectual'" (Žižek 2002: 19). Such disciplinary controversies are never simply arguments about the most accurate method or theoretical framework. They are always also contests for the public perception of what occupies the culturally privileged position of science. As sociologist Thomas Gieryn notes, "[t]he cultural space of science is a vessel of authority, but what it holds inside can only be known after the contest ends, when trust and credibility have been located here but not there" (Gieryn 1999: 15). For Gieryn, scientific authority is not "an always-already-there feature of social life, like Mount Everest. Epistemic authority does not exist as an omnipresent ether, but rather is enacted as people debate (and ultimately decide) where to locate the legitimate jurisdiction over natural facts" (Gieryn 1999: 15). This contest over the privileged position of science becomes manifest, for example, when Wilson laments the fact that "astonishingly, the high culture of Western civilization exists largely apart from the natural sciences. In the United States intellectuals are virtually defined as those who work in the prevailing mode of the social sciences and humanities" (Wilson 1978: 203). The heavy-handed rhetorical load carried by the word "astonishingly" associates sociobiologists and other scientists (presumably sympathetic to the sociobiological project) with reason and clear-mindedness and those working in the humanities and social sciences with silly and dangerous naivety.

As the previous quote suggests, language plays a crucial role in the construction of scientific authority in public debates over science. Professional and popular discourse, however, differ in their choice of textual strategies. As geologist Scott L. Montgomery observes, professional science relies on technical discourse, the very presence of which functions as a claim to cultural authority by producing "an authoritative, insider voice" (Montgomery 1996: 4). This voice, he argues, enacts an asymmetrical and hierarchical distribution of power by making us feel "unprivileged, closed out, ignorant, and—most of all—innocent" as if "excluded from a certain grown-up world of truth and truth-telling" (Montgomery 1996: 2). Since popular science texts cannot deploy technical discourse, they typically substitute a claim to insider knowledge of that discourse for the sense of exclusion such discourse would actually generate. This is the case with Wilson's text as well. Rather than trying to provide a digested version of technical discourse, *On Human Nature* abandons professional language and privileges what it represents as shared cultural vocabulary. It also refuses the "repression of the individual writer" that Montgomery sees as constitutive of technical discourse, favoring instead a highly subjective voice (Montgomery 1996: 24).

While such a step into the polyphonic realm of culture might indicate a failure to claim epistemic privilege, the way in which Wilson employs the idea of an epic invokes a new kind of authority—that of foundational cultural mythology superior to any of the major religions. The text's subjective voice produces a sense of intimacy between the implied author and the implied reader, thereby situating the text outside the exclusive realm of science. Yet Wilson's vision of the unifying power of the evolutionary epic positions him, the author, as the privileged interpreter of both science and culture. As an insider of both realms, he holds the rare skill of translating the cryptically encoded truths of science into the everyday language of human experience. In other words, Wilson strives to secure scientific authority by claiming cultural authority. Even when Wilson invites his readers, as Varghese and Abraham (2004: 212–13) note, to be critical of his claims, the epic frame-work he invokes undermines this caution. His evocation of the epic, that is, renders potential critiques of his account as the uninformed rant of the uninitiated.

Critiques of sociobiology have tended to attack any suggestion about the possibility of shaping human destinies as a sign of a new eugenics. For example, in their response to C. H. Waddington's (1975) review of Wilson's *Sociobiology*, the Sociobiology Study Group explicitly invokes "sterilization laws and restrictive immigration laws by the United States between 1910 and 1930" and "the eugenics policies which led to the establishment of gas chambers in Nazi Germany" (Allen *et al.* 1975). In *On Human Nature*, Wilson indeed argues that "to chart our destiny means that we must shift from automatic control based on our biological properties to precise steering based on biological knowledge" (Wilson 1978: 6). Such a claim is more than ideo-logical icing on an objective, scientific foundation. While Wilson's claims are fundamentally intertwined with value-laden cultural discourses, the working of ideology in the text is also inseparable from the representation of evolution as a linear, ascending narrative. In this sense, the narrative structure of evolution is also a mechanism of ideology. However, while Wilson's evolutionary narrative is able to evoke foundationality through its causal organization and evocation of the epic, his representation of human nature as immutable produces a sense of motionlessness that contradicts the ascending narrative movement. Despite the narrative pull of Wilson's representation of evolution, such narrative friction haunts the text's claims to foundationality and, by extension, to epistemic privilege.

Evolving ambiguities

This chapter has argued that Darwin's evolutionary narrative is structured on a set of ambiguities. There is the tension between a linear teleology of progress and a much more unpredictable trajectory of evolving interspecies continuity. There is also an implicit investment in origins and destinies in Darwin's texts, and yet a sense that those origins are multiple and beyond reach, and that future is always threatened. It is through these tensions that Darwin's evolutionary vision has given rise to various, mutually contradictory versions of evolutionary narrative. It is also through these tensions that evolution is most fundamentally implicated in cultural politics of change, nostalgia, and futurity. Nevertheless, what emerges from Darwin's texts is a sense of foundationality—a sense that natural selection and sexual selection are productive forces that engender material and conceptual change. This foundationality is characterized by constant forward movement that will not cease at the birth of new species or the extinction of old ones.

In *On Human Nature*, Wilson appropriates the foundational potential of Darwin's evolution narrative, while leaving out the inherent unpredictability that characterizes Darwin's interspecies continuity. For Wilson, evolution appears as a linear procession of adaptations that has, in the case of *Homo sapiens*, only recently reached its end. However, Wilson's text carries its own ambiguities, such as the conceptual tension between past movement and present stability in human evolution, or between its methodological circularity and the linear evolutionary procession it imagines. Wilson's teleological narrative is also possible only through retrospection—the main mechanism that produces linearity—and an almost total exclusion of the always chaotic moment of evolutionary present.

What is most striking about Wilson's book, though, is its appropriation of the potential for extensive epistemic claims and mythologizing in Darwin's writing in order to promote a controversial new discipline. Through the idea of evolutionary epic, Wilson builds a rhetorically appealing case of epistemic privilege. This strategy invests the sociobiological evolutionary narrative with a meta-level significance, which in turn acts as rhetorical evidence for the evolutionary processes and rationales that the text imagines. The idea of epic also provides the adaptationist narrative engine with a sense of futurity, thus resolving the textual dilemma that Wilson's claim of adaptive halt in human evolution poses for his evolutionary narrative. Furthermore, while Darwin sought to avoid a clash between science and religion, Wilson rushes head-on to the evolution–creation controversy that has occupied US culture for almost a century. Crucially, however, much of the criticism that Wilson's sociobiology has faced is not antievolutionist but rather a different reading of the relationship between evolution, teleology, and epistemology. We shall return to this point in the next chapter.

These similarities and differences between Darwin and Wilson suggest that the ambivalence and imaginative richness of Darwin's treatment of origins and destiny has made evolutionary biology the groundbreaking project it is today. However, it has also left evolution open to continuous debate and strategic appropriation. The continuities and disruptions between Darwin and Wilson also suggest that foundationality and teleology are not mutually dependent, and that foundationality can in fact be explicitly anti-teleological. At the same time, Wilson's text demonstrates that foundationality can be used to enforce and naturalize teleological reasoning. These shifting associations of foundationality, teleology and transformation provide an opportunity for various readings and revisions of evolutionary processes as culturally meaningful acts. While the facts about nature, gender, and sexuality that such narratives produce are not equally true statements about the empirical world, the evolutionary narratives that they are part of extend beyond the empiricist logic of correct versus incorrect.

Finally, Wilson's adaptationism ties his textual practice to cultural politics of gender. For example, Wilson's account of early human societies privileges male agency, as he tends to focus on male innovation, such as "cooperation during hunting" (Wilson 1978: 85). The feminist reception of Wilson's and other sociobiologists' texts registers this gender bias. In *Science and Gender*, for example, Ruth Bleier (1984) critiques sociobiology for ignoring the role of women in human evolution as gatherers and cultural innovators. But Wilson's sociobiology is also gendered on a more fundamental level, as his adaptationism depends on the reproduction of adaptive traits. This highlighting of reproduction brings a whole range of gendered and sexual implications to his evolutionary narrative. As we shall see in Chapters 3, 4, and 5, it is through this reproductive narrative logic that sociobiology and evolutionary psychology become ultimately entangled with ideas of gender, sexuality, and reproduction.

Note

1 See Dorothy Nelkin and M. Susan Lindee ([1996] 2004: 38–57) for a discussion of DNA as a sacred text in popular culture.
2 For discussion of evolutionary theory before Darwin, see Ruse (2006: 28–63) or Bowler (2007: 30–78).
3 These included, among others, Morgan ([1972] 2001), Reed (1978), Gould ([1981] 1996), Bleier (1984), and Lewontin *et al.* (1984).
4 For an overview of the introduction and development of Darwinian theory, see Ruse (2006) or Larson ([2004] 2006).
5 Similarly, Dov Ospovat argues that "Darwin still believed nature was programmed to achieve certain general ends" at the time of the publication of the *Origin* (Ospovat 1980: 193).
6 Before Darwin, changes in the natural world over history had been addressed by Erasmus Darwin, Jean Baptiste Lamarck, Georges Cuvier, Karl Ernst von Baer, Richard Owen, Charles Lyell, and Robert Chambers, among others. For discussion, see Larson ([2004] 2006) or Ruse (2006).

7 The phrase is from Tennyson's poem "In Memoriam A. H. H." ([1908] 2003), first published in 1850. The poem mourns the death of Tennyson's friend Arthur Henry Hallam, who died in 1833. The poem touches upon key Victorian themes, including the longstanding debate about religion and natural history.

8 In progressivist interpretations, evolution is not just a series of mindless narrative events driven by the logic of incessant change, but instead proceeds from simplicity to complexity, from a tentative innovation (a rudimentary structure enabling the processing of light) to a highly developed trait (the complex eye of the fly). See Ruse (2006) for discussion.

9 For an overview of the controversy, see Segerstråle (2000) or Bethell (2001). For an early insightful commentary on the rhetoric used in the controversy, see Albury (1980).

2 Narrative variation and the changing meanings of movement

In 2005, the Federal District Court in Harrisburg, Pennsylvania, became the scene for *Kitzmiller* v. *Dover Area School District*, a widely publicized lawsuit that challenged the school district's decision to include "intelligent design" (ID) in the public school science curriculum. The case was part of a long series of judicial encounters between evolutionists and antievolutionists dating back to the 1920s. The *Kitzmiller* case also marked a climactic point in a more recent development in the evolution–creation controversy. As the news media repeatedly emphasized, judge John E. Jones III was the first justice to rule on the scientific status of ID, the new "scientific" branch of creationism that insisted that some phenomena in nature are simply too complex to be products of random selection. Not surprisingly, the media coverage of the trial and the popular literature that it engendered highlighted the historical significance of the *Kitzmiller* trial. Portrayed as "a warfare of ideas" dating "way back to the adoption of the First Amendment" (Weiss 2005) or even "all the way back to the Middle Ages" (Wallis 2005), the evolution–creation controversy was understood as having reached a critical moment that would "define (or redefine) for decades just what children are taught about where we come from" (Humes 2008: xii).

Edward O. Wilson's *On Human Nature* explored in the previous chapter was written, published, and read within the evolution–creation controversy. As we saw in that chapter, Wilson's invocation of foundationality appropriates and reinforces culturally resonant ideas of origins and destiny, gender and sexuality, as well as freedom and determinism. In particular, Wilson's appropriation of the epic engages in the long-standing battle over the cultural status of evolution as a foundational narrative. The evolutionary epic Wilson envisions robs religion of cosmological insight, reducing it into "a product of the brain's evolution" (Wilson 1978: 201). For Wilson, religiosity is a neurologically controlled "readiness to be indoctrinated" that may have "evolved through the selection of clans competing one against the other" (Wilson 1978: 184). Philosopher Michael Ruse captures the metaphysical scope of Wilson's evolutionary epic when he notes that "Wilson may be right that he has shucked the literal apocalyptic commitments of his childhood, but if he is not committed to a postmillennial theology, I do not know who is" (Ruse 2006: 213).

Religion is perhaps the most significant cultural discourse against which evolutionary narratives have been, and continue to be, articulated. This is the case especially in the United States, where a large Christian fundamentalist population has a considerable cultural influence. However, religious discourse is also a factor that many British authors of popular science are acutely aware of and seek to negotiate, as popular writing on science—especially popular science books—are typically published and distributed across the English-speaking world. Significantly, cases like the Dover trial have engendered an almost unanimous rejection of all forms of creationist belief from scientists. As Tom Bethell (2001) points out in his reassessment of the sociobiology controversy, evolutionary biologists agree almost point to point in their critique of creationism and intelligent design. Yet there are considerable differences among evolutionists in how they understand the scope and nature of evolutionary processes. While evolutionists agree that evolution is ultimately a material process, they disagree on how we should understand materialism, and what its relationship to religion or spirituality is. There is, then, a constant tension between the joint defense of evolutionary science by evolutionists, and the struggle among evolutionists over what should be the epistemically superior branch of evolutionary thought. Even when evolutionists attack each other's ideas, they all seek to secure the cultural status of science as an epistemic authority. This is a complicated rhetorical task, which writers tackle with a range of strategies.

This chapter traces similarities and differences among supporters of evolutionary theory in their visions of evolutionary change and the role of humanity. Whereas the previous chapter traced a historical movement from Darwin to sociobiology, this chapter examines the variation of evolutionary narratives in the past two decades. This analysis helps us understand the multiplicity of the Darwinian narrative, especially its structural potential and limitations. These narrative parameters in turn shed light on the popular appeal of particular models of evolutionary narration. As in the previous chapter, very little of what follows is actually about gender and sexuality. Yet the seemingly nongendered issues examined here have fundamental implications for the cultural understanding of gender and sexuality. In particular, contemporary evolutionists' revisions of the ideas of transformation, stability, and foundationality affect the ways in which gender can (or cannot) be understood as an object of political debate.

The chapter begins with a brief review of the evolution–creation debate, the cultural context of evolutionary arguments directed to the general public. It then turns to two contemporary versions of Wilson's evolutionary epic, so-called Epic of Evolution texts that seek spiritual fulfillment in the evolutionary narrative and philosophical naturalist texts that posit matter as the ultimate level of all explanation. Despite the stark difference in their tone, these two approaches to evolution both appropriate the ambiguous foundational potential in Darwin's writing. These two sets of texts demonstrate that claims of foundationality are implicated in claims for epistemic authority, and that the two are often articulated through nationalistically coded discourses.

The second half of the chapter turns to two novels in order to highlight the cultural reception of different versions of evolutionary foundationality. John Darnton's *Neanderthal* represents human evolution as a teleological narrative of progress while invoking both American nationalist discourse and religious imagery. Will Self's *Great Apes*, by contrast, depicts human evolution as a distinctly British narrative of decline and fall by positioning chimpanzees as the evolutionarily superior species that rules modern London. Crucially, the way in which both novels imagine progress and failure within a strictly binary narrative dynamic is suggestive of structural and cultural constraints set on evolutionary narratives.

The evolution–creation controversy

Darwinian evolutionary theory was widely accepted in most of Europe by the mid-twentieth century. Although there were critical voices that continued questioning natural selection and the relatedness of humans and apes, the modern synthesis of Darwinian evolutionary theory and Mendelian genetics in the 1930s gave evolutionary theory the status of a proper science. This development resulted in the inclusion of evolutionary theory in school science curricula across the continent. In the US, however, the situation was very different. The early history of the new nation had left a strong conservative Protestant legacy. This religious tradition sought to read the Bible literally and thus obviate all acts of interpretation between people and their God.

This religious legacy gave rise to the Fundamentalist movement at the turn of the twentieth century. The movement gained increasing prominence in the 1910s, giving rise to growing attempts to remove evolution from school curricula. In 1925, these attempts led to the introduction of the Butler Act, which outlawed the teaching of evolution in publically funded institutions in Tennessee. The Act was challenged in the famous Scopes "Monkey" Trial later in the same year. In the trial, the criminal court of Tennessee ruled over the inclusion of evolution in a science class by a young teacher, John Scopes, who had been encouraged by the American Civil Liberties Union ACLU to test the law. The court chose to omit the question of the scientific validity of evolutionary theory and instead gave Scopes a fine for breaking the state law. While Scopes avoided the sentence due to a technicality, this also prevented the ACLU from taking the case to the Supreme Court of Tennessee. As a result, the Butler Act was upheld, and was soon followed by similar legislation in other states.[1]

In 1968, the Supreme Court of the United States finally outlawed the antievolution laws of Arkansas, Mississippi, Louisiana, and Tennessee that banned the teaching of evolution in public school science classes. The ruling forced creationists to develop alternative strategies for challenging evolution. These included the removal of evolution from the core requirements for state-level exams, the addition of a note in science textbooks stating that evolution is "just a theory," and the promotion of an alternative "creation science." In 1987, the Supreme Court decided on *Edwards* v. *Aguillard*, a lawsuit that challenged the Louisiana law that mandated "equal time" for "creation science" and evolution in public school science classes. The case resulted in yet another blow for the creationists, as the Supreme Court ruled that "creation science" was not a genuine science and thus did not belong in science classes.

Rather than give up the fight over the cultural authority associated with science, creationists developed the new "discipline" of intelligent design (ID), which claimed to satisfy the standards of science while avoiding any direct mention of religion or the Bible. Throughout the 1990s, ID gained increasing popularity as a seemingly serious rival of evolution. While the Dover school district was the first one to include ID in their science curriculum, other state education boards and local school boards were engaged in similar attempts. Instead of rejecting the whole discipline of evolutionary biology, ID advocates have attacked specific instances of evolutionary theory, claiming that nature is full of highly complex phenomena that must have been designed by a superior intelligence. These claims have been rejected by evolutionists, who challenge the assumption that complexity can emerge only from intention, and that a current gap in the evolutionary record means that there will always be a gap that evolution cannot explain.[2]

ID is more sophisticated in its use of scientific arguments than most strands of creationism. Yet it is not unique in its appeal to scientific authority. As cultural anthropologist Christopher P. Tourney (1991) demonstrates, creationist groups willing to invoke "scientific" arguments in support of their views have been more successful in persuading non-committed audiences than openly anti-science sects or organizations. Similarly, anthropologists J. Patrick Gray and Linda D. Wolfe suggest that invoking "the prestige of science" is necessary for any project in order to "gain the adherence of the majority of Americans" (Gray and Wolfe 1982: 587). Indeed, sociologist Kathleen E. Jenkins (2007) shows that many fundamentalist evangelical leaders appropriate the language of molecular biology—especially the iconic status of DNA as the "sacred" code of life—in order to argue for genetic essentialism (such as the patriarchal division of gender roles as written in the genome) while combating scientific naturalism. This suggests that religious and scientific discourses are not alien to one another. Rather, they provide rhetorical resources that both evolutionists and creationists seek to deploy. This discursive appropriation is not merely an American phenomenon, but organizes British accounts of evolutionary origins and destinies as well.

The meaning of matter

In the popular science literature published in the 1990s, there emerge two sets of texts that address evolution and religious concerns while appropriating Wilson's understanding of evolution as a foundational narrative. The first group of texts is often called Epic of Evolution. It includes such books as

Brian Swimme and Thomas Berry's *The Universe Story* (1992), Ursula Goodenough's *The Sacred Depths of Nature* (1998), Loyal Rue's *Everybody's Story* (2000), and Eric Chaisson's *Epic of Evolution* (2005), which all evoke the evolutionary epic as a source of ethical guidance and spiritual fulfillment. The second set of texts is often referred to as philosophical naturalism. Such books as Richard Dawkins' *River Out of Eden* ([1995] 1996) and *The God Delusion* (2008), Daniel Dennett's *Darwin's Dangerous Idea* (1996) and *Breaking the Spell* (2007), Malcolm Potts and Roger Short's *Ever Since Adam and Eve* (1999), and Victor J. Stenger's *God: The Failed Hypothesis* (2008) all advocate an openly atheist approach that rejects spirituality and emphasizes the intellectual superiority of evolutionary thought. While these two revisions of evolutionary foundationality differ radically in their attitude toward religion, they both stand in stark contrast to the more conciliatory approaches proposed by scientists like Stephen Jay Gould (1999) or Francisco J. Ayala (2007). Unlike Epic of Evolution or philosophical naturalism, Gould's and Ayala's *methodological naturalism* proposes that the natural world should be examined through the methods of science while leaving spiritual questions of human existence to religion and philosophy.[3] Proponents of Epic of Evolution and philosophical naturalism, by contrast, reject the assumption that the spiritual and the material occupy different epistemic realms.[4]

Inspired by Wilson's idea of evolution as an all-encompassing origins story, the Epic of Evolution approach has produced a number of books, conferences, and even pioneering college courses since the mid-1990s.[5] The goal of the Epic of Evolution is to bring together evolutionary and cosmological sciences in order "to engage the larger philosophical and religious communities in an ambitious attempt to understand truly who we are, whence we came, and where we are headed as wise, ethical human beings," as Eric Chaisson envisions in *Epic of Evolution* (Chaisson 2005: 441). While finding its source of spiritual fulfillment and moral guidance in scientific discoveries, the Epic is nevertheless friendly toward religion and, according to proponent Loyal Rue, does not "rule out belief in the reality of a personal deity" (Rue 2000: 132). Some proponents go still further, evoking, like Brian Swimme and Thomas Berry in *The Universe Story*, "the human soul" (Swimme and Berry 1992: 242, 246), or praising, like Ursula Goodenough in *The Sacred Depths of Nature*, "the deep wisdom embedded" in traditional religions (Goodenough 1998: 173). The Epic of Evolution, then, departs from both methodological naturalism's restricted scope and philosophical naturalists' belief that materiality excludes spirituality. For advocates of the Epic of Evolution, science is neither a mere tool for understanding natural laws, as it is for methodological naturalists, nor a proof of the superiority of materialism over spirituality, as it is for philosophical naturalists. In Epic of Evolution texts, science is a search for philosophical and spiritual insight in the natural world.

Philosophical naturalist texts invest the narrative of cosmic evolution with a very different meaning. Instead of the mythic role assigned to evolution in

On Human Nature, philosophical naturalism builds on Wilson's bold assertion that evolutionary theory can "explain traditional religion, its chief competitor, as a wholly material phenomenon" with the consequence that "[t]heology is not likely to survive as an independent intellectual discipline" (Wilson 1978: 192). Philosophical naturalist texts often explicitly attack religion, as when Daniel Dennett accuses devout believers of being "irresponsible" (Dennett 2007: 50) and having "self-indulgent fantasies" (Dennett 2007: 256), Richard Dawkins describes religion as "a force for evil in the world" (Dawkins 2008: 324), or Victor Stenger portrays God as an equal to "Bigfoot, the Abominable Snowman, and the Loch Ness Monster" (Stenger 2008: 245) and those in faith as "gullible" (Stenger 2008: 248). At the same time, philosophical naturalist texts typically represent evolution as an epistemically superior account of any phenomenon—a recurrent theme, for example, in Richard Dawkins' or Daniel Dennett's texts. Rhetorical scholar Thomas M. Lessl calls such assertions of epistemic privilege "gnostic scientism" in which scientific knowledge occupies a privileged position as "the singular experience of consciousness that enjoys immunity from the deterministic powers of natural evolution" (Lessl 2002: 143). Giving the argument a further twist, Slavoj Žižek describes philosophical naturalist claims as "obscurantist sprouts" that "function as what Louis Althusser would have called a 'spontaneous ideology' of scientists themselves, as a spiritual supplement to the predominant reductionist-proceduralist attitude of 'only what can be precisely defined and measured counts'" (Žižek 2002: 23). Such claims of authority and privilege typically attack both religion and postmodernism, a rhetorical strategy that suggests a likeness between the two largely incompatible projects, associating them both with a dangerous lack of clarity of vision.

Despite the obvious differences between Epic of Evolution and philosophical naturalist texts, the evolutionary narratives produced in these texts share considerable structural similarities. Science studies scholar and science writer Jon Turney (2001) argues that the discoveries of cosmology and evolutionary biology provide a narrative pattern that can be appropriated to promote radically different values in popular science. As a result, he maintains, "contemporary popular science, while offering interpretations of how things are in the natural world, becomes another arena for disputing about the human future" (Turney 2001: 226). As Turney shows, there is not simply a foundational and anti-foundational version of the evolutionary narrative, but the very idea of foundationality arising from the eon-spanning scope and retrospective linearity of evolution can be revised and appropriated in many ways. While Epic of Evolution and philosophical naturalist texts rely on a similar reading of the narrative logic of Darwinian evolution, they rework its foundational potential in different ways. The narrative strategies they deploy demonstrate sensitivity to the cultural values their imagined audiences presumably savor. This becomes evident in how Epic of Evolution and philosophical naturalist texts appropriate not only the epic potential in Darwinian evolution but also various nonscientific cultural narratives.

This reworking of foundational potential is not just a matter of adding a moral or ideological layer on a scientific core. Turney maintains that if the cosmic–evolutionary narrative "is to be transformed from a simple ordered narrative into a story, it needs a moral. And the moral needs to be supplied by the story-teller" (Turney 2001: 229). Contrary to Turney, I consider the presence of a moral as produced, to large extent, by the organizing logic of the evolutionary narrative rather than by a cultural "outside." This is visible, for example, in how both Epic of Evolution and philosophical naturalist texts appropriate the irreversibility characteristic of evolution. While invoking the idea of evolution as constant forward movement, many texts invest this irreversible procession with goal-mindedness, as when Swimme and Berry discuss the horse's "sacred journey into its destiny" (Swimme and Berry 1992: 137) or Dawkins ([1995] 1996) invokes the expulsion from the biblical Eden as an allegory of evolution. At the same time, a moral is not an intrinsic part of narrative structure. Rather, it arises from the ideologically charged interface between the structural tendencies of the evolutionary narrative and the cultural discourses in which evolutionary narratives are embedded.

Such appropriation of structural and discursive elements rewrites evolution as not only foundational but also teleological in Epic of Evolution and philosophical naturalist texts. The sense of logical necessity is reinforced by the positioning of humans at the endpoint of the evolutionary narrative. In philosophical naturalism, the materialist explanation of human behavior typically appears as the ultimate concern of evolutionary science and the culmination of evolution itself. Similarly, Epic of Evolution texts encourage us to fully appreciate the "bonded community" of all life forms (Swimme and Berry 1992: 35) in order "to embrace our cosmic heritage, to make fuller use of our human potential" (Chaisson 2005: 442). Both approaches produce narratives of the human condition, thereby reflecting and reinforcing cultural assumptions about the uniqueness of humans. The role of humanity in the universe is, of course, one of the most sensitive issues in the whole evolution–creation controversy. This significance underlies, for example, the burning of the "evolution of man" mural in Dover High School, an event that often appears as a key narrative moment in popular accounts of the *Kitzmiller* trial (see Humes 2008 or Slack 2007). In this sense, the implicit anthropocentrism of Epic of Evolution and philosophical naturalist texts could be read as negotiating cultural expectations. These texts equate the irreversibility of evolutionary change with the retrospective recognition of modern humanity as the endpoint of the evolutionary narrative. They also evoke the popular understanding of narrative as movement toward the end, and of the end as a culmination of that narrative.[6]

Another example of this balancing between structural constraints and cultural assumptions is the highlighting of progress in both Epic of Evolution and philosophical naturalist texts. Although the structural ambiguity of the Darwinian evolutionary narrative is what allows assumptions of progress in the first place, these assumptions also resonate with the ideology of progress, which is typically represented as rooted in the Enlightenment and thus as intrinsic to modern science and Western—and especially American—mentality. Epic of Evolution texts tend to locate progress in natural processes, which, "pervaded by inherent tendencies toward fulfillment of their potential" (Swimme and Berry 1992: 53), exemplify "the astonishing property of emergence" (Goodenough 1998: 30). By emphasizing purpose and perfection, Epic of Evolution texts rewrite progress as ethically just, a strategy that takes issue with the antievolutionist portrayal of evolution as inherently nihilistic. Echoing Darwin's representation of natural selection as morally guided, Swimme and Berry maintain that there is "wild wisdom at the heart of the universe story" (Swimme and Berry 1992: 89), especially in the principle of natural selection, which, as "a tendency toward interrelatedness" (Swimme and Berry 1992: 133), promotes the "survival of the most cooperative" (Swimme and Berry 1992: 123) rather than the most competitive. At the same time, many Epic of Evolution texts contrast this ethically sound progress with the false idea of progress represented by "the myth of Wonderland" that will be "if only we continue ... the ever-increasing exploitation of the Earth" (Swimme and Berry 1992: 218). While arising from the structural potential of Darwinian evolution, this narrative of progress appropriates the cultural significance placed on advancement in order to articulate environmentalist and spiritual concerns. Intriguingly, the narrative that emerges—humans disturbing the mythic "equilibrium" (Rue 2000: 6) of the evolutionary "paradise" (Swimme and Berry 1992: 140)—is both structurally parallel and discursively similar to the biblical account of the Fall.

Philosophical naturalist accounts, by contrast, often evoke the familiar narrative of scientific progress in which the endlessly accelerating chain of scientific discoveries, typically understood as the culmination of evolution, stands for the progress of Western culture—which, in the context of US cultural politics, is often interpreted as specifically American progress. Through this set of associations, evolutionary progress appears as synonymous with scientific and technological advance, on the one hand, and epistemic privilege, on the other. While philosophical naturalist texts tend to represent nature without mysticism—consider, for example, Daniel Dennett's (1996: 48–60) description of natural selection as an algorithm—the victorious advances of science often acquire a mythic dimension. This is evident, for instance, in Dawkins' *River Out of Eden*. Imagining the picture of man and woman carried by a space capsule "deliberately sent on an eternal outward journey among the stars," Dawkins asserts: "[T]his couple is not Adam and Eve, and the message engraved beneath their graceful forms is an altogether more worthy testament to our life explosion than anything in Genesis" (Dawkins [1995] 1996: 187). Furthermore, philosophical naturalist texts often contrast progress with religious faith, as when Stenger portrays religion as "primitive, archaic images from the childhood of humanity" (Stenger 2008: 40) fundamentally "inimical to human progress" (Stenger 2008: 248).

Even those philosophical naturalist accounts that express concern about ecological balance represent sociobiological evolutionary theory as the only possible remedy, thereby echoing Wilson's vision of sociobiology as the antidote to the failures of modern life. Of special concern for Malcolm Potts and Roger Short's *Ever Since Adam and Eve*, for example, is our perceived failure to acknowledge our inner (highly gendered) proclivities. This failure has rendered us "hunter-gatherer women and men lost in a concrete jungle" and "spear-throwing warriors with fingers on the nuclear button" (Potts and Short 1999: 312). Such a simultaneous appropriation of the foundational potential in evolution and the cultural prestige associated with the narrative of scientific advance gives the evolutionary narrative mythic significance. Unlike the biblical epic, however, this philosophical naturalist evolutionary narrative does not depict the Fall as resulting from the acquisition of (sexual) knowledge but from a failure to gain such knowledge, to recognize "the animal within us" (the title of Potts and Short's final chapter). This appeal to evolutionary foundationality not only rejects the biblical origins story. It also dismisses as ridiculous feminist and queer scholars' understanding of gender and sexuality as always implicated in cultural politics. This dismissal is reinforced by such linguistic choices as the reference to "cultural fundamentalists" (Potts and Short 1999: 74) or the portrayal of a constructionist view of gender as "silly" (Potts and Short 1999: 72).

The biggest leap from a slight structural reworking of the epic potential in Darwinian evolution to a full-fledged deployment of nonscientific cultural assumptions takes place when Epic of Evolution and philosophical naturalist texts invoke a paradigm shift. This is a common rhetorical strategy, through which writers seek to produce epistemic authority. Many Epic of Evolution texts represent their cherished epic as "the most important intellectual endeavor of the new millennium" (Rue 2000: 129) while positioning themselves "at a remarkable time when truly fundamental issues can be addressed" (Chaisson 2005: 11–12). Philosophical naturalist texts, on the other hand, evoke a shift from a false religious worldview to a true Darwinian view, as in the case of Dawkins' ([1995] 1996) symbolic river out of religious folly. Such claims are given additional weight through a series of analogies that evoke the cultural prestige of high art. Epic of Evolution texts tell us that "[p]atterns of gene expression are to organisms as melodies and harmonies are to sonatas" (Goodenough 1998: 58), that "the Milky Way expresses its inner depths in Emily Dickinson's poetry" (Swimme and Berry 1992: 38), and that Walt Whitman is "a space the Milky Way fashioned to feel its own grandeur" (Swimme and Berry 1992: 40).[7] Similarly, Dawkins' philosophical naturalist evolutionary narrative is not only "a grander and incomparably more ancient epic" (Dawkins [1995] 1996: 66) but "maybe even more poetically moving" (Dawkins [1995] 1996: 38) than any religious creation narrative, and Potts and Short's claims about human nature are supposedly validated by canonical literary works (Shakespeare and Dickens are their favorites). In such a narrative framework, evolutionary science—instead of, say, literary criticism or musicology—becomes the privileged means of "decipher[ing] the story" of human history and culture (Chaisson 2005: 433).

Notwithstanding their differences, both Epic of Evolution and philosophical naturalist texts engage in strikingly similar narrative politics in order to argue for the cultural significance of their particular view of evolution. While seizing and reworking the narrative potential in Darwinian evolution, they invoke and endorse cultural narratives of human destiny, nationalistically coded progress, and ultimate meaning. Although Epic of Evolution and philosophical naturalist texts build on Darwin's understanding of natural selection as the fundamental mechanism of evolutionary change, the resemblance between these contemporary texts and Darwin's inherently ambiguous narrative is in many ways distant. Where Darwin's evolutionary narrative is permeated by Victorian curiosity, controversy, and imperialist imagination, the foundational narratives that Epic of Evolution and philosophical naturalist texts produce are strategically engaged in contemporary cultural politics, especially the American debate over religion and science.

These similarities between Epic of Evolution and philosophical naturalism demonstrate that the ways in which foundationality and transformation can be conceived and appropriated are not endless. The prominence of particular textual strategies—teleological trajectories, valorization of scientific advance, the juxtaposition of progress and collapse—suggests that there are certain ideologies of movement that resonate with cultural expectations better than others. Both Epic of Evolution and philosophical naturalist texts rely on continuous forward movement. They are also underwritten by a sense of danger— the possibility of extinction or the destruction of the planet—that produces the end of the evolutionary trajectory, the ethically (Epic of Evolution) or scientifically (philosophical naturalism) aware humanity, as a desired narrative outcome. This does not mean that no contemporary version of evolutionary narrative challenges this ideology of movement. For example, Niles Eldredge and Stephen Jay Gould's (1972) theory of "punctuated equilibrium" critiques the model of continuous adaptive change known as phyletic gradualism by suggesting that the evolution of species is characterized by long episodes of stasis, which are interrupted by sudden periods of rapid change. While emphasizing the stability of species, punctuated equilibrium nevertheless recognizes the continuity of change both on the level of the organism and the level of global evolutionary change. This suggests that the narrative ideology of movement is not tied to a specific cultural politics or scientific approach, as the accounts it engenders may be mutually contradictory.

What is curious about the ideology of movement is that evolutionary narratives often include characteristics that are immutable. As we know from sociobiological discourse, gender and sexuality are prime examples of stable entities. Another such entity is national integrity, which underlies both Darwin's texts and philosophical naturalism. One explanation of this phenomenon is that these stable entities function as narrative anchors. That is, they operate as constants that help fix the narrative trajectory of continuous change. While they are not a necessary part of the narrative logic of evolution, these largely frozen entities allow evolutionary narratives to extend in time and space without collapsing into uncertainty. Evolutionary narratives that portray gender, sexuality, nation, and human nature as fixed entities suggest that our destinies may be secured, that things will not fall apart in a cosmic web of unpredictability. I return to the question of the stability of gender and sexuality in the next three chapters.

Replaying the evolutionary tape

What we saw in Epic of Evolution and philosophical naturalist texts could be described as a balancing act between the essential (if ambiguous) properties of evolution by natural selection, the texts' epistemic commitments, and the expectations of the texts' non-specialist audiences. In this sense, the two sets of texts are suggestive of the narrative politics involved in writing about evolution to non-specialist audiences immersed in the polemics of the evolution–creation controversy. Cultural expectations are not, however, a fixed set of assumptions that popular science texts can contain or control. Rather, cultural expectations are diverse, historically contingent, and in constant flux. While evolutionists may choose the bits and pieces that suit their purposes—a particular view of progress, for instance—their account of the evolutionary narrative may yet face an ambivalent reception. While reaffirming deeply felt cultural values—the significance placed on "roots" in determining who we are as social, cultural, and biological beings, for example—Epic of Evolution and philosophical naturalist texts also challenge the long-cherished belief in human (and especially American) uniqueness at the center of the cosmos: we may be the final, triumphal product of evolution but we are still a product of evolution. While resonating with popular fantasies of discovering our true origins (and, by implication, our unfolding destinies), evolutionary narratives suggest the possibility that human existence may simply lack a higher purpose.

Having traced a series of mutations of the Darwinian evolutionary narrative in popular scientific discourse, I want to take a look at the ways in which evolution is interpreted as a foundational narrative outside the educative and argumentative genres of popular science. Such an analysis helps us understand the challenges that popularizations and defenses of evolutionary theory need to tackle in order to appeal to non-specialist audiences. The rest of this chapter explores two novels, John Darnton's popular scientific thriller *Neanderthal* ([1996] 1997) and Will Self's satiric novel *Great Apes* (1997) in order to examine the ways in which evolutionary narratives are implicated in cultural discourses of gender, sexuality, religion, and nationalism. Both novels tackle the scientifically contested and culturally sensitive connection between human evolution and the evolution of closely related species: Neanderthals in Darnton's book and chimpanzees in Self's book. In this sense, the books' treatment of the human/nonhuman boundary anticipates the publication of the Neanderthal genome in 2010 and the publication of the chimpanzee genome in 2005. Both of these genome projects focused on the evolutionary distance between modern humans and the studied species.

Darnton's *Neanderthal* is a typical and formulaic representative of its genre. The plot opens with two paleontologists and former lovers, Matt Mattison and Susan Arnot, discovering the disappearance of Jerome Kellicut, their former teacher at Harvard, during a secret one-man expedition in the Pamirs. Summoned by the head of the Institute for Prehistoric Research, Matt and Susan learn that Kellicut had found evidence that Neanderthals still exist. Matt and Susan embark on a rescue expedition with three others. Searching for shelter in a mountain cave during a blizzard, the rescuers are attacked by a band of Neanderthals and try to escape through the tunnel network intersecting the mountain, where they discover that their hominid enemies are brain eaters. At the other side of the mountain, Matt and Susan are rescued by another band of Neanderthals, who take them down to a temperate valley. There they encounter a peaceful hominid tribe and the missing Kellicut and learn that the hominids possess a special faculty for "remote viewing," an ability to enter others' minds and see through their eyes. This revelation confirms their suspicion that the Institute is a government agency masquerading as an independent scientific institution, and that it is seeking more than just scientific discovery. As narrative conventions require, the novel ends with Susan's imprisonment by the evil hominids, Matt's ingenious rescue operation, the defeat of the cave-dwelling renegades by the peaceful valley dwellers, and the final failure of the government forces to reach the Neanderthals and thus utilize their special faculty for military purposes. Consistent with genre expectations, the novel closes with the reunited lovers sipping champagne on a flight back to the US.

Darnton's novel relies on an understanding of the narrative dynamic of evolution similar to many Epic of Evolution and philosophical naturalist texts. Like these texts, *Neanderthal* not only emphasizes our connectedness to other species but also places us at the endpoint of evolutionary history, as when Matt, observing the cave-dwelling Neanderthals, imagines himself as "witnessing the birth of civilization, the moment in which our ancestors turned from the brutish existence of solitary apes to the splendor and rigors of community and industry" (Darnton [1996] 1997: 290). In her analysis of late nineteenth and early twentieth-century paleoanthropological texts, anthropologist Misia Landau (1991) argues that such texts often follow the narrative structure of a heroic tale in which the early hominid—the symbolic representative of the birth of humanity—has to pass a set of tests. According to Landau, these tests function "to raise the hero from a primitive human state to civilization. Given that this was the objective right from the start, the achievement of civilization is the hero's final triumph" (Landau 1991: 11). This is the logic that organizes *Neanderthal* as well: the prehistoric hominid is never just a prehistoric hominid but represents modern humanity in making. As in Epic of Evolution and philosophical naturalist texts, this portrayal of *Homo sapiens* as the endpoint of the evolutionary narrative is further reinforced by the discursive positioning of evolution as our primary origins story and evolutionary theory as the epistemically privileged interpreter of those origins. Since "[e]very tribe has its own central myth," religious beliefs are, by necessity, plural and relativistic (Darnton [1996] 1997: 192). At the same time, science emerges as unified and objective through the portrayal of both protagonists—our primary points of association—as devoted scientists and of their relationship as moving toward the (academic and romantic) reconciliation suggested at the outset by the genre.

The novel's anthropocentrism, however, does not end here. In *Neanderthal*, we are not just the final fruit of blind evolutionary processes. If evolution started from scratch, the text suggests, it would follow the same steps. In *Wonderful Life*, Stephen Jay Gould challenges such a "march of progress" (Gould [1989] 1990: 31), arguing that a "replaying" of "life's tape" would produce an altogether different evolutionary narrative (Gould [1989] 1990: 48). According to Gould:

> Each step proceeds for cause, but no finale can be specified at the start, and none would ever occur a second time in the same way, because any pathway proceeds through thousands of improbable stages. Alter any early event, ever so slightly and without apparent importance at the time, and evolution cascades into a radically different channel.
>
> (Gould [1989] 1990: 51)

When picturing the cave-dwelling Neanderthals as representing "the moment in which our ancestors turned from the brutish existence of solitary apes" to civilization, Matt is invoking the predestined march of progress challenged by Gould. Crucially, Matt is not suggesting a mere analogy between human prehistory and the scene he is observing. Since his Neanderthals are not actual prehistoric ancestors of humans but a contemporary species descended from the prehistoric Neanderthals, Matt is also articulating the popular assumption that if we replayed humanity's evolutionary tape, we would end up with human culture. The text's explicit references to destiny reinforce this effect, as when Kellicut asserts that the peaceful valley-dwelling hominids "weren't meant to prevail" (Darnton [1996] 1997: 313) while the cave-dwelling renegades "were clearly marked by destiny as the future of the species" (Darnton [1996] 1997: 369).

At the same time, the assumption about humanity's inevitability is somewhat undermined by the novel's contradictory treatment of Neanderthals both as our now-extinct evolutionary competitors and as a transitional species leading to modern humans. This ambiguous positioning of Neanderthals as both similar and different, both "us" and "them," is a common theme in the popular accounts of Neanderthals in the media. These accounts usually focus on the conundrum of "their" disappearance and "our" survival, as in the tellingly titled "Last of the Neanderthals" published in *National Geographic Magazine* (Hall 2008) or "Twilight of the Neandertals" published in *Scientific American* (Wong 2009). Both articles are informed by the same question:

How did modern humans gain an edge over Neanderthals in the struggle for diminishing resources? While both writers emphasize Neanderthals' physical and cognitive skills, the underlying narrative is that of human success and Neanderthal failure. Yet the texts acknowledge that there is no consensus whether Neanderthals are "a separate species" or "an archaic variant of our own species" (Wong 2009: 35).

Darnton's *Neanderthal* spells out this ambiguity. At one point in the novel, a senior paleontologist at the Institute contemplates the psychological and cultural motivation behind the long-standing portrayal of the Neanderthal as a brain eater:

> [W]e need something to separate us from him to put us back up on our pedestal. We need to transform him into a beast. What better way to do so than to accuse him of violating the most pernicious taboo imaginable, committing the most heinous crime, the symbol of everything that places us above others on this horrible continuum of struggling savages—eating your own kind?
>
> (Darnton [1996] 1997: 247)

The novel articulates this desire to differentiate through its narrative treatment of the valley dweller/cave dweller/human triad. While the novel portrays us as the heirs of the brain-eating, cave-dwelling renegades, these hominids are symbolically distanced through detailed, cinematic descriptions of the humans' horror and repulsion when witnessing the cave-dwellers' cruelty. This human–hominid difference is further highlighted by the novel's representation of the renegades as social misfits who have left the "tranquil rhythms" of the secluded valley (Darnton [1996] 1997: 203). Our true origins, then, are not in the violence of the renegades but in the beauty and harmony of the valley.

In order to establish this harmonious past for humankind, the novel needs to portray the valley dwellers as noble beings. Kellicut, for example, sees them as "kindred beings who exist on a higher plane" (Darnton [1996] 1997: 199) where "communication occurs in its purest form" (Darnton [1996] 1997: 201) and "everyone is truly equal" (Darnton [1996] 1997: 199). They also demonstrate exceptional heroism, sacrificing themselves for their comrades. Such a valorization of the valley dwellers reaffirms human superiority and uniqueness. Erasing what Darwin famously called "the indelible stamp" of our "lowly origin"—our descent from a simple life form—the novel imagines a noble genealogy equal to those of traditional epics (Darwin [1879] 2004: 689). Accordingly, the valley appears as a "primordial world" (Darnton [1996] 1997: 314) where the humans witness "the beginning of time" (Darnton [1996] 1997: 140). This portrayal of the valley as lacking a past contradicts the novel's explicit reference to the valley dwellers as the descendants of those prehistoric Neanderthals who survived the harsh battles between *Homo sapiens* and *Homo neanderthalensis*. It also necessitates the representation of the valley-dwelling hominids as if they had not evolved for the last 30,000 years. Giving voice to

[handwritten margin notes:] like a Hobbes vs Rousseau debate!

[handwritten note at bottom:] → paints modern humanity in a 'natural' state, as if we are basically the same as cave people

this fallacy, Matt imagines the valley as "a real-life prehistoric laboratory" (Darnton [1996] 1997: 234), and when touching a Neanderthal hand, experiences "a pulsating thrill so strong it seemed a throwback to some earlier age, as if his genetic core was ignited by that spark of contact" (Darnton [1996] 1997: 183). This oxymoron of "current prehistory" results in the ideologically loaded representation of modern humans as evolutionary adults and the contemporary hominids as evolutionary children. Contradicting the very real existence of Neanderthals that has been the driving force of the whole narrative, the final chapter pictures Susan thinking of the hominids as our "older brothers and sisters who died in infancy" (Darnton [1996] 1997: 389).

This representation of the valley as the noble birthplace of humanity is given symbolic resonance through the evocation of biblical imagery. To an extent, this cultural resonance functions to make evolution more palatable for those non-committed readers who are yet reluctant to abandon the idea of noble origins and higher destiny. When Matt regains consciousness after the escape through the tunnels, he associates his first sight of light behind the trees with "biblical etchings of the forest primeval at the beginning of creation" (Darnton [1996] 1997: 177). The valley itself is the biblical garden, as the title of Part II, "Eden," spells out, and as Kellicut explains to the protagonists: "You have found Eden itself, the great garden of paradise, before Adam and Eve's transgression" (Darnton [1996] 1997: 202). Moreover, relying on culturally familiar anthropocentrism, the novel portrays the human protagonists rather than the hominids as "Adam and Eve strolling through the lush prelapsarian garden" (Darnton [1996] 1997: 213). Such an allegory locates human origins with the modern American couple while rendering evolution a series of inherently meaningful steps in the service of a higher order. At the same time, the discursive framing of the valley dwellers as prelapsarian—as "innocent, naive, trusting beings" (Darnton [1996] 1997: 202)—also provides another means of highlighting the cultural, temporal, and spatial distance between "us" and "them." Our roots are indeed in prelapsarian goodness and innocence, but what long roots those are.

As with the biblical Eden, however, the timeless harmony of the valley is bound to come to an end. This narrative inevitability is partly due to the text's evocation of biblical imagery, which implies a fall from grace. At the same time, this sense of inevitability is a result of the very logic of the evolutionary narrative: if we are the endpoint of evolution, as the text assumes, the story needs to get going so that it can get to us. As in the biblical account of Eden, transgression functions as the primary impetus of change (and thus of narrative) and is represented as already latent in the seemingly perfect harmony. If the valley is the birthplace of the future renegades, Matt observes, then "it's a contrived Eden. There's a darkness at the center" (Darnton [1996] 1997: 261). The novel accommodates this symbolic fall from grace by portraying the renegades as representing a higher evolutionary stage: the valley-dwellers are "eons behind" (Darnton [1996] 1997: 188) their cave-dwelling contemporaries, who embody, in Kellicut's words, the historical moment when "[a] species is

> greater, but not more morey – tension between
> the goodness of innocence and greatness

we are
freer now...

going to reinvent itself, shed its old self like a used skin and become something greater, something more advanced" (Darnton [1996] 1997: 313). Through this narrative strategy, the text implies that what might seem like a fall in fact signifies ascent to something higher. Since the endpoint of the evolutionary narrative is already determined, the renegades appear as a necessary stage on the path toward humanity. This also functions to justify their violent behavior, as suggested by Kellicut's speculation on killing as "a necessary way station on the road to civilization" (Darnton [1996] 1997: 260). In this, the novel reflects what philosopher and sociobiology proponent Patricia Williams asserts to be the main lesson of sociobiology: "In the model built by the doctrines of original sin, conflict is a sign of bondage. In nature, conflict is a sign of freedom, with freer creatures being more conflicted" (Williams 2001: 146). There is a promise of salvation in this view of human nature, as it confirms the association of conflict with progress. This emphasis on freedom as a sign of true civilization and violence as inevitable also resonates comfortably and reassuringly with American cultural sensibilities.

These narrative negotiations also underlie the novel's portrayal of American national progress. In her analysis of Robin Cook's medical thriller *Chromosome 6*, Priscilla Wald identifies a narrative logic that posits the novel's genetically engineered hominids as an "evolutionary mirror" through which the text establishes a prehistoric genealogy while reconfirming modern humanity's assumed superiority (Wald 2005: 212–16). In this discursive frame, the hominids'

> biological evolution is displaced onto their rhetorical evolution into a national identity ('American') and, in turn, that national identity is represented in evolutionary terms. America represents the epitome of civilization and, therefore, of human evolution: implicitly, the triumph over the human instinct to violence.
>
> (Wald 2005: 215)

Darnton's *Neanderthal* demonstrates a similar narrative logic. Through its reworking of evolutionary, biblical, and national narratives, the text establishes the two Neanderthal tribes as worthy points of origin for modern humans. Whereas the valley dwellers represent the inherent nobility of human origins, the cave dwellers reinforce popular assumptions of human cultural progress through their own progress toward humanity. At the same time, the cave dwellers' abhorred violence projects true progress on human evolution, thus celebrating modern humanity's "triumph over the human instinct to violence" (Wald 2005: 215). That this true evolutionary narrative is also a national narrative is suggested by the fact that the renegades are driven to their final destruction in an avalanche by Bruce Springsteen's "Born in the USA" playing from the belly of the primitive Trojan horse designed by Matt, the novel's thoroughly American hero.

Such textual politics demonstrate that assumptions of foundationality are central to the popular understanding of evolution in American culture. It also implies that despite the heat of the evolution–creation controversy, discourses of evolution and religion are not opposed but rather mutually embedded in contemporary America. While echoing and expanding the strategic appropriation of religious discourse characteristic of Epic of Evolution and philosophical naturalist texts, *Neanderthal* does not simply reject religion, like philosophical naturalism, but rather builds on the unresolved cultural tension between the evolutionary and biblical narratives. While Darnton's text grants evolution epistemic privilege, it also acknowledges and even endorses the centrality of religious discourse in American culture. At the same time, the repeated invocation of the idea of progress casts human evolution as an implicitly American narrative in which the logic of evolutionary change functions as the symbolic guarantee of future national progress. If we replayed the evolutionary tape, modern humanity—understood as the contemporary United States through the figures of the all-American hero and heroine— would securely emerge. Furthermore, this American narrative is an individualistic narrative of bravery and endurance, as Matt and Susan stand in opposition to the always suspect institutional forces represented by the Institute.

This intertwining of evolutionary, religious, and national narratives also makes the text's gender politics safe and familiar. For example, the narrative of progress that organizes *Neanderthal* renders the potential reawakening of the earlier romantic relationship between Susan and Kellicut a narrative impossibility. While their past affair functions as a necessary impediment in the relationship between Matt and Susan and thus as a constitutive complication of the novel's narrative trajectory, Kellicut's refusal to leave the valley associates him firmly with the evolutionary past, casting him as a figure antithetical to narrative futurity. The novel also invokes the possibility of intimacy between Susan and one of the male hominids, thereby echoing the fear and fascination that research on modern humans' and Neanderthals' short coexistence has engendered in popular culture. Indeed, the whole Neanderthal genome project was informed by the question whether modern humans and Neanderthals interbred. In the aftermath of the 2010 publication of the Neanderthal genome, this question was rewritten as one of "a Neanderthal-modern human one-night stand" versus "thousands of interspecies assignations" in popular scientific discourse (Than 2010). While Susan's emotional and sensual reaching toward the Neanderthals has obvious queer potential, the association of the hominids with the past and the representation of evolution as a narrative of progress render such boundary-crossing romance an unlikely narrative outcome. Instead, the biblical framing of Susan and Matt as Adam and Eve implies that they are destined to move on and multiply—symbolically if not literally—thereby releasing the future from the grip of the past. As Adam and Eve, Matt and Susan are the safely white, straight, and successful American couple that holds the future of humanity in their hands. *Neanderthal*'s intertwining of evolutionary, religious, nationalistic, and romantic narratives, then, reinforces hopes of progress associated with science in contemporary culture while soothing cultural anxieties arising from the materiality and unpredictability of evolutionary processes.

Reversed destinies

If Darnton's *Neanderthal* plays with the familiar idea of progress as inherently American, and, by extension, of Americanness as the organizing principle of the narrative of progress, Will Self's *Great Apes* provides an altogether different vision of evolution, progress, and national history. Situating the evolutionary narrative in the context of British national history, Self's novel sheds further light on the tensions inherent in the idea of progress and transformation. Published in 1997, Self's novel imagines an alternative course for human evolution as it traces the faith of artist Simon Dykes, who wakes up after an alcohol and drug-filled night in a world ruled by chimpanzees. Simon is diagnosed with a rare delusion that makes him believe that he is human and that the humans are the evolutionarily successful species. He is thus faced with a painful process of rehabilitation and readjustment to the chimpanzee society. With the help of the "anti-psychiatrist" Dr Zack Busner (Self 1997: 28), self-styled "eminent natural philosopher" (Self 1997: 377) and "the noted authority on chimpanzee nature" (Self 1997: 92), Simon gradually learns to accept his "chimpunity." In Self's novel, human descent appears as a failed narrative, a mere flat trajectory that has led to "the human evolutionary cul-de-sac" (Self 1997: x) and impending extinction. Chimpanzee evolution, by contrast, has reached the proper narrative climax marked by cognitive and intellectual development, technological progress, and economic and cultural growth. While such flipping of narrative destinies highlights the possibility of failure intrinsic to evolution, evolutionary history nevertheless appears as a foundational narrative of modern culture.

The evolutionary narrative that *Great Apes* constructs is a distinctly British narrative built on the narrative model of decline and fall often evoked in the context of the post-World War II disintegration of the British Empire.[8] It also plays with the idea of devolution central to Victorian debates about evolution.[9] Set in London in the late 1990s, the novel examines contemporary metropolitan life, especially the pretensions, ambitions, and failures it associates with the London art scene, night clubs, public hospitals, transportation system, and the masses crowding the streets. Inhabited by anthropomorphic chimpanzees, London stands for the culmination of chimpanzee evolution and thus as a reminder of the corresponding failure of human evolution. Self's chimp world is distinctly satirical. With Jonathan Swift's *Gulliver's Travels* as its most obvious predecessor, *Great Apes* posits Simon's human delusion as a critique of "the condition of modern chimpunity" (Self 1997: 220) conceived "in the manner of a satirical trope" (Self 1997: 404). The chief means through which this satirical effect is produced is the portrayal of the chimpanzee Britain as a near replica of the extratextual human world: *Great Apes* sports a chimp Freud, chimp Shakespeare, and chimp Jane Goodall—

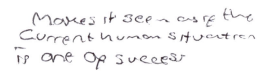

Makes it seen as if the
Current human situation
is one of success

the famous observer of humans in Gombe valley—as well as cultural products like *The Planet of the Humans* movies or the chimp Maugham's *Of Chimpanzee Bondage*. Even the ethnic groups of the human world have their primate counterparts, as the bonobos, the unprivileged chimpanzee minority, are revealed to be "the race of chimpanzees who inhabit Africa" (Self 1997: 340). Through this set of parallels, Self's text also parodies the common representation of evolutionary theory as a distinctly British invention and of Darwin as a national hero, as well as undermines Darwin's portrayal of "[t]he remarkable success of the English as colonists" as a sign of evolutionary advance (Darwin [1879] 2004: 167).

The novel's evolutionary narrative is poised between the alternative narratives of success and failure, both of which are familiar from long-standing cultural debates about the state of humanity. On the one hand, the novel evokes the common assumption of humanity as the culmination of the evolutionary narrative, implicit in Epic of Evolution and philosophical naturalist texts and explicit in Darnton' *Neanderthal*, while projecting its organizing logic and key events onto chimpanzee history. On the other hand, the novel's account of human evolution appropriates the narrative of inevitable fall by identifying a point in the past when the narrative subject (humanity) failed to produce a key narrative development (an adaptation) that would have pushed the story toward its climax. A renowned chimpanzee philosopher at Oxford University locates this fateful moment in the emergence of chimpanzee "signage," which is understood in the novel as the sole efficient form of communication: "Unlike the chimpanzee, whose signage competence had evolved over two million years of continuous selection, determined by brain-sign interaction, the human had become bogged down in a perverse and clamorous sound garden" by falsely privileging speech over gesticulation (Self 1997: 298). While the narratives of chimpanzee progress and human failure rely on a similar idea of origins from which the narrative is understood as emerging, they diverge through the reconstructed moment of narrative crisis. This parallel deployment of progress and failure emphasizes the tension between the potential fulfillment and the potential failure to reach that fulfillment that characterizes narrative in general and evolutionary narrative in particular. By making the unpredictability of future failure and success its central theme, *Great Apes* dramatizes the structural tension that gives narrative its sense of movement and suspense. At the same time, the novel also operates as a literal demonstration of the retrospective nature of our knowledge of evolutionary change. It is only from the perspective of the present that the past appears as secured.

The novel's representation of chimpanzees as (almost) human echoes cultural anxieties about the close genetic relationship between humans and chimpanzees. Like the difference between modern humans and Neanderthals, the human–chimpanzee species boundary is often conceived in hierarchical terms in popular culture. This tendency is evident, for example, in *Time* magazine's article "What Makes Us Different?" (Lemonick 2006). Consistent with the

→ intelligence and resourcefulness as being the defining human success traits,

long tradition of representing chimpanzees as failing humans, the piece evokes modern humans' "brainpower to outthink and outdo our closest relatives on the tree of life" through our "ability to speak and write and read, to compose symphonies, paint masterpieces and delve into the molecular biology that makes us what we are" (Lemonick 2006). In *Great Apes*, this tension between the narratives of human and chimpanzee evolution is reinforced through the chimpanzee civilization's striking familiarity and radical difference. While the cultural and historical detail of the chimpanzee world corresponds with the human civilization Simon imagines (and the reader knows), there is a set of social practices relating to "those most chimpmane activities, sex and violence" that stands out as alien to modern humanity (Self 1997: 379). It would obviously be highly unlikely that the blind processes of natural selection produced a chimpanzee replica of the human world that also included some characteristically chimpanzee traits. Rather than a logical failure in Self's text, however, this contradiction serves a satirical function as it draws attention to those human social practices and cultural phenomena that the chimpanzees find appalling or alien.

Makes it seem as if humanity shows progress

Significantly, this textual strategy ends up denaturalizing patterns of human behavior that are often taken as grounded in biology. For example, by portraying the chimp society as convinced of the superiority of polygamy, *Great Apes* casts human sexual practices as a "perverse adhesion to the organizing principle of monogamy" (Self 1997: ix) and jealousy as an unusual and pathological case of "emotional distress" (Self 1997: 340). Similarly, the text undermines the obviousness of the incest taboo that has occupied such a central place in both anthropology and sociobiology. In the novel's chimpanzee society, Freud is known for demonstrating "the destructive emotional effect of a biological alpha not mating his daughter" (Self 1997: 142). Furthermore, the human avoidance of touch and curtailing of aggressive behavior become defamiliarized through the text's emphasis on bodily contact. While grooming is represented as the social glue that precedes and follows every chimpanzee encounter and conflict, violence appears as constitutive of both society and the individual's sense of belonging, as suggested by "the reassuring blows" (Self 1997: 149) delivered on a subordinate by his or her superior and Simon's consort's doubt "that her mother really cared about her, so infrequently did she attack" (Self 1997: 141). In such a world, Simon is considered seriously ill because the core chimp patterns of "vocalisation, body signage, grooming, presenting, all of these are grossly impaired, if not absent altogether" (Self 1997: 123).

Suggests however there is nothing inherently better about human customs, if it is all about what 'works' for a society

This sense of defamiliarization, however, has its limits. Despite the fact that the narrative of chimpanzee progress unsettles deep-seated cultural beliefs about the obviousness and superiority of what is commonly understood as essentially human characteristics, the human and chimpanzee evolutionary narratives appear as imprisoned in a binary logic. By reversing human and chimpanzee destinies, *Great Apes* ties the two diverging evolutionary narratives into a dynamic of success and failure in which the progress of one

→ takes away the immutability of custom as often suggested in appeals to nature e.g. marriage between Am and Jw, has been for millennia...

narrative necessitates the regress of the other. Significantly, this binary logic also underlies other evolutionary trajectories in the text such as the reversed destinies of the lap ponies kept as pets and the horse-sized dogs used for riding. This sense of mutual dependency makes the specific endpoints of the chimp and human evolutionary narratives seem like necessary outcomes of the organizing logic of the two intertwined narratives. Both the human and chimpanzee ways of arranging sexual and social relationships, that is, appear as if dictated by the parallel trajectories of the two species' evolutionary histories. Moreover, this binary logic implies that there are only two possible models of behavior. While such assumptions about destiny do not necessarily mark a narrative failure—*Great Apes* is, after all, a satire of contemporary issues—they echo and consequently reproduce the popular idea of evolution as goal minded. Even the satirical tone cannot quite undermine this sense of imperative necessity. This becomes evident toward the end of the novel, when Simon rediscovers his instinctive knowledge about the practices of mating, presenting, and displaying. His human delusion, then, is ultimately just that: a delusion that is replaced by his growing acceptance of what he re-recognizes as reality.

What increases this sense of foundationality is the way in which *Great Apes* connects these intertwined narratives of progress and failure to the immediately recognizable narrative of cultural crisis. The novel repeatedly portrays the chimp modernity as potentially destructive to itself, thereby echoing the popular view of modern life as having failed to respect our most basic biological needs. Familiar from reactionary social commentaries in general and sociobiological dystopias in particular, the crisis narrative is manifest, for example, when Zack Busner ponders whether "the current woeful condition of chimpunity is a maladaptive response to overcrowding, to the loss of our natural lifestyles" (Self 1997: 37). The novel's satirical tone produces a critical distance to such commentary, positioning the text implicitly within the long tradition of laments on the destructive influences of the present on the nostalgic, authentic past. Yet such comments attest to the popular appeal of the idea of evolution as a matter of ascent or descent. As Gillian Beer argues, this tension between ascent and descent constitutes a fundamental ambiguity in Darwin's writing, as "[t]he optimistic 'progressive' reading of development can never expunge that other insistence that extinction is more probable than progress" (Beer [1983] 2000: 6). This idea of evolution as a mechanism of both rise and fall has shaped debates about evolution and the human condition throughout the nineteenth and twentieth century. Self's novel's constant evocation of the crisis narrative—the inevitability of the end of ascent—thus engenders a sense of predestination, suggesting that modern society will get what it deserves. This portrayal of modernity, of course, resonates with the apocalyptic visions that often accompanied popular speculations about the approaching millennium in the late 1990s, the historical context within which Self's novel appeared.

Great Apes, then, demonstrates that unpredictability and foundationality are not mutually exclusive characteristics. Whereas Darnton's *Neanderthal* invests evolution with assumptions of foundationality and teleology, *Great Apes* suggests that teleology does not follow automatically from foundationality. Self's novel produces a sense of foundational movement notwithstanding the text's insistence that evolution is never fully secured and appears obvious only in retrospect. In *Great Apes*, the parallel and oppositional treatment of the human and chimpanzee evolutionary narratives engenders a sense of irreversible procession toward two mutually dependent futures. This logic of movement suggests that foundationality is an effect of the structural dynamic that the text builds among different evolutionary narratives. This indicates that the sense of linearity and inevitability that often characterizes popular narratives of evolution does not arise from any single feature in evolutionary thought. Rather than a matter of either form or content, foundationality is evasive, disappearing and reappearing somewhere between narrative structure and cultural context, in a zone characterized by conceptual friction.

Self's text also suggests that the connection between evolution and epistemology is complex and subtle. While explicit claims about the epistemic privilege of evolutionary theory are largely absent in *Great Apes*, the text nevertheless posits evolution as the primary framework within which chimpanzee and human nature is interpreted. Epistemic claims, then, are implicated in the novel's narrative framework rather than explicitly articulated. Even though the novel's satirical tone refuses a full commitment to the sociobiological discourse of genetic advantage and reproductive strategies, the binary logic that organizes the parallel evolutionary trajectories produces a sense of fundamental revelation. This textual ambiguity demonstrates that chance, destiny, teleology, and foundationality are not neatly locatable features but rather intersect and intertwine with a range of cultural discourses in various intertextual encounters.

Ambiguities and anxieties

This chapter has traced textual strategies through which evolutionary narratives have produced a sense of foundationality in the cultural contexts of the past two decades. I have argued that the narrative production of foundationality is intertwined with the rhetorical production of scientific fact, and that both scientific authority and the sense of fundamental truth are implicated in discourses of gender, sexuality, nation, and religion. While evolutionary narratives vary in their detail, emphasis, and narrative structure, and can be appropriated for diverse purposes and for different audiences, there are structural constraints to such processes of narrative mutation. In other words, there is a point where evolutionary narratives cease to be *evolutionary* narratives. Analyzing ideas of transformation and foundationality in evolutionary narratives thus becomes a project of tracing cultural negotiations over what counts as the correct interpretation of Darwin's theory, on the one hand, and the evolutionary narrative's structural resistance to rewriting, on the other.

The ambiguity about teleology and foundationality at the heart of Darwinian evolution has played a significant part in the controversies surrounding evolution in the late twentieth century. While the ambivalence and imaginative richness of Darwin's treatment of origins and destiny has made evolutionary biology the groundbreaking project it is today, it has also left evolution open to continuous debate and strategic appropriation. Crucially, much of the criticism sociobiology and evolutionary psychology have faced is not antievolutionist but rather a different reading of the relationship between transformation, teleology, and epistemology. The same is true about the evolution–creation controversy, which has never been simply a clash between proponents and opponents of evolution. As the comparison of Epic of Evolution and philosophical naturalist texts suggests, involved in the controversy are also different evolutionist interpretations of the reach and implications of evolutionary explanation.

The evocation of foundationality serves a range of ideological purposes in different texts and historical contexts, as is suggested by Epic of Evolution texts' encouragement of spirituality and rejection of genetic opportunism, and philosophical naturalism's rejection of spirituality and endorsement of the language of innate competitiveness. Moreover, the ways in which claims of foundationality are implicated in particular cultural discourses vary, as is evident in the texts' invocations of nationalist discourse. Epic of Evolution insists on fairness and inherent progress, both commonly perceived as characteristic of the true American spirit and the role of the US in the world, whereas philosophical naturalism justifies its atheism through an appeal to progress and freedom, the hailed and contested object of the American cultural imagination. Similarly, *Neanderthal* demonstrates how the potential for foundationality can be used to buttress a nationalistically coded American narrative of progress, while *Great Apes* provides a critical rereading of the British version of the narrative of decline and fall.

Notwithstanding this textual variation, all the texts build on what could be called a shared set of core structural characteristics. They all emphasize narrative continuity, which they imagine as spanning from the distant past into the future, driven by that most Darwinian invention, natural selection. They also tend to conceptualize evolution in terms of ascent or descent, indicating the difficulty of imagining evolution as steady narrative movement not based on a clearly distinct series of crucial key adaptations (or their fatal failure). This tendency to highlight moments of evolutionary change also underlies distinctly non-adaptationist accounts such as Niles Eldredge and Stephen Jay Gould's punctuated equilibrium. The texts studied here imagine key moments in evolution through the binary of descent versus ascent, as when some philosophical naturalist texts, following Wilson's epic, portray human evolution as a march of progress toward a point in the future where the story will either collapse or reach ever higher destinies. Furthermore, this narrative dynamic often posits different evolutionary trajectories as mutually dependent, so that one species' success depends on the failure of another. The way in which both *Neanderthal* and *Great Apes* invoke this dual dynamic suggests that such antagonism resonates with cultural expectations. Finally, many of the texts emphasize the linearity that arises from retrospection while deemphasizing the chaotic multitude and seeming lack of direction that characterizes the moment of evolutionary change. Through this emphasis on retrospective linearity, the texts studied in this chapter omit the myriad alternative routes that evolution might have traveled, or indeed did travel.

While all these texts play with ideas of foundationality and epistemic privilege, the connection between the two concepts is complex and cannot be reduced to any particular feature of the narrative. Rather, the association between foundationality and authority arises from the conjoiner of the narrative logic of continuous, irreversible change and discourses of religion, nationalism, gender, and sexuality. Similarly, teleology—the belief in preset historical trajectories—and foundationality—the belief that events have fundamental consequences for ontological or epistemological questions that far exceed those events—may or may not be implicated in each other. Whereas *Neanderthal* and some Epic of Evolution texts evoke both, *Great Apes* embraces foundationality but rejects teleology. Nor are chance and teleology necessarily mutually exclusive, as even those evolutionary narratives that subscribe to a teleology—*Neanderthal* is a case in point—tend to explicitly thematize the importance of chance in evolution. While evolution is always, to an extent, foundational, neither teleology nor unpredictability is a universal characteristic of evolutionary narratives.

We have also seen that there are crucial fractures in evolutionary narratives, such as Darnton's inherently contradictory representation of the Neanderthals as both prehistoric and contemporary, or Self's portrayal of chimpanzee evolution as having led to an implausible mixture of distinctly human and distinctly chimpanzee characteristics. These inconsistencies are indicative of cultural anxieties surrounding evolutionary narratives as well as evolutionary narratives' inability to accommodate them, and they may thus help us understand the wider cultural politics in which evolutionary narratives participate. These contradictions also shed light on the textual strategies involved in negotiating the gap between theoretical models and cultural expectations, as when *Neanderthal* and Epic of Evolution texts tone down their claims about the epistemic superiority of materialism by evoking biblical and spiritual discourse.

The above texts share only a partial resemblance with the evolutionary narrative Darwin sketched in *The Origin of Species* and *The Descent of Man*. Twentieth- and twenty-first-century versions of the Darwinian evolutionary narrative have extended, developed, and revised its narrative potential into models that, in some cases, have proved theoretically innovative. One such case is the discovery of the DNA molecule and the role of the gene as the unit of selection in the second half of the twentieth century. In the next chapter, I turn to the emergence of the molecular imaginary and its impact on the evolutionary imagination. In the texts discussed above, gender appears as either implicitly embedded in the cultural discourses evoked by evolutionary narratives, or as a constitutive but largely untheorized part of the narrative of the evolution of a species. The next chapter asks what happens to ideas of gender, transformation, and foundationality when cultural assumptions about sexual order are projected onto genes and chromosomes, and evolution becomes written as a dichotomously gendered molecular narrative. In particular, the chapter interrogates the ways in which gender becomes increasingly seen as the driving force of the evolutionary narrative.

Notes

1 For a historical account of the US trials contesting the role of evolution in public school science curricula, see Scott (2005). For a discussion of the constitutional issues involved in the controversy, see Irons (2007). For a very helpful overview of different religious positions on evolution, see Scott (2005).
2 For scientific critique of the claims made by ID proponents, see Scott (2005), Ayala (2007), or Antolin and Herbers (2001). See Baird (2000) for an analysis of the philosophical and historical roots of the "just a theory" argument.
3 Ayala insists that religion focuses on "questions of value, meaning, and purpose that are forever beyond science's scope" (Ayala 2007: 163). Gould imagines "a respectful, even loving, concordat between the magisteria of science and religion" (Gould 1999: 9) provided that religious views "no longer dictate the nature of factual conclusions" (Gould 1999: 9) and philosophical naturalists refrain from "claim[ing] higher insight into moral truth from any superior knowledge of the world's empirical constitution" (Gould 1999: 9–10). As many have pointed out, such a curtailing of the realm of religion is unlikely to appeal to anyone other than the most liberal and secular church members. For a critique, see Bowler (2007: 11) or Ruse (2006: 205–6).
4 This distinction between methodological and philosophical naturalism is a common one in debates about the relationship between science and religion. It is invoked, for example, by Eugenie C. Scott (2005), Francisco J. Ayala (2007), and Kenneth R. Miller (2007)—expert witness for the plaintiffs in the *Kitzmiller* trial. Many defenders of evolution reject the outspoken atheism of philosophical naturalism as harmful to the status of evolution in the public opinion.
5 Conferences include the 1996 IRAS conference "The Epic of Evolution" on Star Island, New Hampshire, the 1997 AAAS conference "The Epic of Evolution" in Chicago, and the 2008 "The Evolutionary Epic" conference in Waianae, Oahu, Hawaii. For pedagogical experiments, see Russell Merle Genet (1998), who outlines the Epic of Evolution course taught at Northern Arizona University in 1998.
6 See Roof (1996: 1–40) for discussion of the significance of the ending in narrative.
7 This representation of nature as poetry (or as the true author of all poetry) can also be seen as an extension of what Patricia Waugh identifies as "the scriptoral metaphor" that gained popularity in the second half of the twentieth century, producing scientific accounts that imagined "nature as an autopoetic invention, a writing which writes itself, liberated both from demiurges and authors" (Waugh 2005: 243).
8 See Cannadine (2005) for a discussion of different models of British history.
9 See Beer ([1983] 2000) or Young (1985).

3 The gendered politics of genetic discourse

Darwin's understanding of the macro-level mechanism of evolutionary change, natural selection, has been fundamental to the development of modern evolutionary theory and remains its organizing premise still today. Darwin could, however, offer only an educated guess as to the micro-level transmission of traits from parent to offspring. Based on speculation rather than observation, his theory of *pangenesis* proposed a unit of inheritance called *gemmule*. Gemmules were minute particles that travelled from all parts of a parent's body to the reproductive organs, and thus they reflected both inherited characteristics and characteristics acquired during the parent's life time (Darwin [1868] 2010). The assumption that habits and exercise affect the reproductive particles, which are then passed on to subsequent generations, came from the famous French naturalist Jean-Baptiste Lamarck. It was not until the Moravian monk Gregor Mendel's experiments on inheritance in peas became known around 1900 that the idea of inheritance as a combination of the parents' unblended traits replaced prevalent assumptions of blended traits and Lamarckian cultivation. The introduction of population genetics (the study of genetic variation within populations) in the aftermath of World War I gave rise to the so-called "modern synthesis" in the 1930s and 1940s. The synthesis connected the insights of Mendelian genetics with those of Darwinian evolutionary theory, thereby giving rise to theoretical advances in both genetics and evolutionary biology.

In today's accounts of the history of the biological sciences, the discovery of the structure of the DNA molecule is typically represented as the key moment in the march toward modern genomics. As *Time* magazine declared on the fiftieth anniversary of the publication of Francis Crick and James Watson's famous article "A Structure for Deoxyribose Nucleic Acid" in 1953, it was "a discovery that in the half-century since has transformed science, medicine and much of modern life—though the full impact has yet to be felt" (Lemonick 2003). While Watson and Crick were the official discoverers of the DNA structure, their model relied on X-ray crystallography by Maurice Wilkins and Rosalind Franklin. In the following decades, this work on DNA gave rise to a plentitude of research initiatives on genes and their cellular and chromosomal environment. This has generated an increasing flow of money into anything genetics related, as well as a cultural hype surrounding the very image of the DNA double helix. An object of popular fascination, the double helix appears today almost anywhere from business logos to television commercials to personalized jewelry.[1] With the introduction of genomics (the study of the totality of genes in an organism) in the last decade of the twentieth century, the gene itself has acquired, in rhetorical scholar Elizabeth Shea's words, the status of "a cultural icon that has come to connote fixed realities and the possibility of material origins of the self," thus promising an answer to fundamental metaphysical questions (Shea 2001: 505).[2]

Science studies scholars have viewed this rise of the molecular imaginary as an articulation of larger conceptual developments in the models and practices of professional science, and the cultural debates through which those models and practices take shape. Many scholars have connected the growing interest in the molecular to the development of information technology and communications science during World War II and the ensuing Cold War, thereby viewing the ambitious Human Genome Project (1990–2003) as the ultimate extension of the logic of information science (Franklin 1995; Haraway 1997; Keller 2000; Marchessault 2000; Thacker 2003; Waugh 2005). Conceived as an enormous computer code while literally stored and processed in digital format, the human genome is the emblematic symbol of the postmodern understanding of the body as information. Furthermore, Judith Roof observes that the hype surrounding DNA is itself merely a recent expression of a long-standing fascination with the minuscule. Underlying this focus is the belief that "[s]cience, knowledge, truth exist in the primary and irreducible. Thinking about truth is a process of finding the 'primary,' 'irreducible,' 'basic,' 'original,' 'underlying,' 'first,' 'substances,' 'principles,' and 'causes'" (Roof 2007: 35). This highlighting of the minute is paralleled by the development of ever more powerful visual technologies. Sociologist Nikolas Rose argues that genomics as an intellectual enterprise in fact depends on the development of technologies of visualization, which, together with new methods of processing genetic material, "opened 'the gene' to knowledge and technique at the molecular level" (Rose 2007: 14). As sociologists Kelly Joyce (2006) and Catherine Waldby (1997) observe, technologies of visualization are symptomatic of the privileging of the visual in postmodern culture. These technologies are a major constituent of the attraction of the minute and the interior.

The emergence of the double helix as a cultural icon has also had a major effect on popular representations of evolution. Through the metaphors of "the book of life," "the human library," and "the human archive" associated with DNA, evolutionary history has come to be seen as implicated in the logic of reading, deciphering, translating, preserving, and editing.[3] As Jon Turney observes, the use of such metaphors suggests that "[w]e can learn to read DNA as an archive of the evolutionary past, and a record of the inter-relatedness of all life" (Turney 2001: 235). If DNA is understood as key to our evolutionary past, then evolution, too, is cast in genetic terms. As a result, evolution appears as a genetically coded foundational narrative and the human genome as the record of the production history of that narrative. This has engendered a fundamental entanglement of discourses of genomics and evolution in the popular imagination, making it difficult to talk about one without evoking the other. However, as genomics is concerned with the minute detail of nucleotides and amino acids while evolution tackles prehistoric eons, the two discourses are in many ways incompatible. Not surprisingly, this discursive entanglement tends to generate a giddying fusion of the digital and the fossilized, the observable and the hypothetical, the microscopic and the telescopic.

Furthermore, the rise of the molecular has also engendered a shift of agency from the organism and the species to the genetic and cellular in evolutionary narratives, thus echoing what Rose calls the "molecularization" of life in contemporary science (Rose 2007: 5). Gabriel Gudding (1996) traces this shift back to the introduction of the distinction between genotype—an individual organism's genes—and phenotype—the expression of those genes in the organism's body and behavior—during the first half of the twentieth century. According to Gudding, this distinction between a trait and its expression "eventually led to a sense of the body's fragility and its eventual 'disappearance' as a seat of agency, morality, and identity" with the result that "these three groundings of modernity have been redistributed to the gene, the genotype, and the genome" (Gudding 1996: 525). Whereas Darwin examined natural selection at the level of the individual and the communal, portraying "the social instincts" as having evolved to serve "the good of the community," evolutionary narratives that appeared after the 1976 publication of Richard Dawkins's seminal *The Selfish Gene* have tended to cast the gene as the evolutionary protagonist (Darwin [1879] 2004: 149).

This chapter asks what happens to the foundational potential of the evolutionary narrative and its politics of movement when evolution is written at the molecular level. In particular, the chapter explores the effect of the rise of the molecular imaginary on the representation of gender and sexuality, on the one hand, and the production of epistemic privilege, on the other. The chapter addresses these questions through a close reading of three popular science texts, Dawkins' *The Selfish Gene*, Bryan Sykes' *Adam's Curse*, and Natalie Angier's *Woman*, as well as Simon Mawer's novel *Mendel's Dwarf*. I suggest that the molecular imaginary provides a narrative dynamic that helps resolve some of the fundamental contradictions that have haunted Darwinian evolutionary narratives by providing a structurally coherent narrative agent: the gene. Contrary to the temporary and fragile subjectivities of species and organisms, the gene is imagined as eternal, immutable, and invulnerable. This evocation of the gene turns molecular entities into anthropomorphic and gendered agents, with the result that gender appears as productive of narrative movement and yet as largely immune to change.

Immortal agents

The previous chapters argued that evolutionary narratives need to negotiate a set of ambiguities intrinsic to Darwinian evolution. The texts discussed were poised between teleology and unpredictability as well as movement and stability. While these tensions are indicative of the ambivalent connection between Darwinian evolution and ideology, they ultimately arise from the underlying structure of the evolutionary narrative. At the same time, that narrative structure is both embedded in and shaped by larger cultural debates and discourses that are not exclusively or even primarily about science.

This narrative ambivalence is complicated by a further structural dilemma. Literary scholar H. Porter Abbott (2003) observes that evolution differs from most origins stories in that it relies on two parallel narrative levels. On the one hand, there is the narrative level of species constituted by ever-changing genetic variation within a population, "a succession of averages with no real existence at all" (Abbott 2003: 148). On the other hand, there is the narrative level of organisms concerned with the "little stories of love and death" (Abbott 2003: 147) that is "the activity of entities in the real, empirical world (hamsters, humans)" (Abbott 2003: 148). A narrative account of Darwinian evolution has to convey activity on both of these levels, a task that is further complicated by a "narrative disjunction" between the two (Abbott 2003: 147). The "little stories of love and death" acted out by organisms are only loosely connected to species-level variation, since organisms are not consciously acting for the survival of the species, nor is there any direct link between a single organism's actions and the future of the species. Yet changes in the evolution of species take place precisely because of the constantly changing variation of traits produced by the enormous total of all these little stories. Sometimes these two narrative levels approach each other, as is the case with the rapid evolution of some viruses or bacteria in a few generations. Nevertheless, the disjunction between an individual organism's action and the fate of the species remains. As Abbott points out, evolutionary narratives differ here from alternative origin narratives, such as creationist and intelligent design (ID) accounts, which operate at only one narrative level. Whereas natural selection is a mechanical principle or an elaborate algorithm that does not consciously cause or design anything, God or an "intelligent designer" can function as a proper narrative actor who initiates the events that shape species and organisms in creationist and ID narratives.

The two-level structure of evolutionary narratives enacts a curious dissolution of narrative events and narrative entities. According to Abbott, the distribution of agency across different narrative levels puts Darwinian accounts of evolution in a paradoxical situation, as they need to negotiate

> the difficulty of constructing an explanatory narrative that shows agency but that has to make do with an apparent lack of entities and even an apparent lack of events, without which, of course, there can be no narrative. Yet because natural selection is a way of understanding change over time, which in turn would appear to be a kind of action, it is difficult to find other terms with which to describe it.
>
> (Abbott 2003: 144)

As we saw in the previous chapters, a key evolutionary event (such as an important adaptation) is always a product of historical distance, as there is no way of distinguishing significance from insignificance prior to the act of looking back. The retrospective nature of evolutionary knowledge also necessitates focusing on a single line of descent among the vast plentitude of lineages, since what counts as an event for one species may be a barely noticeable fluke for another. Evolutionary events, then, are ultimately blurry. They emerge from an act of contextualization that renders them local rather than global, and fleeting rather than stable.

This two-level narrative structure also renders the very possibility of a single narrative actor problematic. The most likely candidate for the role of protagonist, the species, is trapped in the same retrospective logic as the concept of narrative event, since species as an entity can be construed only when the process of speciation is already over—and even then only in the abstract. Darwin himself was well aware of this difficulty, as suggested by his remark that species are "artificial combinations made for convenience" (Darwin [1859] 1985: 456) and that "the only distinction between species and well-marked varieties is, that the latter are known, or believed, to be connected at the present day by intermediate gradations, whereas species were formerly thus connected" (Darwin [1859] 1985: 455). The organism does not fare any better as a narrative agent, since an individual life span is no more than an eye blink in the eon-spanning evolutionary narrative, and the actor that moves the narrative needs to be more persistent.

This narrative difficulty is particularly pressing in popular science texts, which seek to appeal to a wide range of audiences and therefore privilege familiar, linear narratives. The Epic of Evolution texts explored in the previous chapter try to solve this structural dilemma by imagining evolution as a narrative of the emergence of life, in which life itself is the protagonist, as Turney observes (Turney 2001: 234–5). However, the portrayal of evolution as a slow but constant progress toward ever greater complexity of life is hardly satisfactory to most authors, as this narrative lacks the drama of danger, struggle, and death associated with evolution (and narrative in general) in the popular imagination. As a narrative protagonist, life is also difficult to conceive since it seems to include everything and yet nothing in particular. Nor does it strictly speaking act.

This chapter demonstrates that by shifting agency to the molecular level popular science texts can engender a rhetorically appealing and coherent evolutionary narrative that meets with the popular expectations of a "good story." Crucially, this narrative strategy also masks the inherent instability of the gene as a scientific concept. Despite the symbolic singularity of the gene, the gene as a biological unit is rather hazy. Biologist Helge Torgersen argues that interdisciplinary projects within genomics often involve a degree of epistemic confusion, as genes can "take on the form of a statistical correlation; of a molecule such as a stretch of DNA carrying special information; of a functional relationship within a network; of a nested set of information contents related to each other in a way yet unknown, etc." depending on the scientist's disciplinary background (Torgersen 2009: 83). In this sense, the gene is, as Donna Haraway puts it, "a phantom object, like and unlike the commodity" (Haraway 1997: 142). This contrast between symbolic unity and conceptual polyvalence invests the gene with considerable persuasive power. Elizabeth Shea (2001) captures this rhetorical appeal in her analysis of the metonymic function of the gene in cultural discourse. Shea suggests that the gene is metonymic in the sense that it "was intended to refer to an abstract concept while figuring that concept as a material reality" (Shea 2001: 512). However, this working of metonymy "has become camouflaged as the material realities of genes have been established by molecular geneticists, endorsed by the human genome project, and celebrated in the popular press" (Shea 2001: 516). The gene is powerful as a rhetorical figure not in spite of but *because* of its inherent ambiguity.

One effect of this figurative function of the gene is the sense of foundationality it produces. In popular discourse, this foundationality often becomes extended from the genes to other minuscule entities inhabiting the interior bodily universe, so that molecular agency becomes distributed across a range of biological entities imagined to occupy the microcosm under the high-tech microscope. For example, popular representations of genes tend to treat DNA (the stuff genes are made of) and genes (segments of that stuff) as interchangeable so that "DNA stands for the gene as its synecdoche, taking over ideas about heredity as well as becoming an immense, vague compendium of 'information' believed to cause everything from alcoholism to thrill seeking" (Roof 2007: 6). There is also often a curious dilution of differences between distinct levels of biological organization, as entities like chromosomes, gametes (egg and sperm), and occasionally even organs are imagined as occupying the same molecular space as genes. As a result, these minuscule yet mutually incompatible entities appear as fundamental actors in the evolutionary narrative. I shall refer to this form of evolutionary narrative simply as the *molecular evolutionary narrative*. While this narrative covers more than the strictly molecular, it receives its persuasive power from the cultural valorization of the molecular.

This molecular narrative populates the inside of our bodies with gendered narrative actors. Such gendered anthropomorphism is, of course, nothing new in the history of Western science. Feminist historian Londa Schiebinger ([1993] 2004), for example, has traced the "discovery" of plant sexuality in eighteenth-century botany, arguing that Linnaeus' classification system projected social values and conventional gender roles on the structure and functions of parts of the plant. With the introduction of plant sexuality, those parts of the plant that were considered male, such as the stamen, had to be described as playing an active role in reproduction, even if they were previously considered passive (Schiebinger [1993] 2004: 21).[4] Anthropologist Emily Martin (1991) has identified a similar phenomenon in late twentieth-century textbooks on human reproduction, which typically depict the sperm as an adventurous male and the egg as either a passive recipient or a dangerous femme fatale. That these anthropomorphic inscriptions are projected onto a fundamental biological level like the cell, Martin argues, "constitutes a powerful move to make them seem so natural as to be beyond alteration" (Martin 1991: 500). The texts analyzed in the following sections take gendered anthropomorphism as their starting point, rewriting evolution as a story about the ambitions and successes of microscopic men and women.

Foundational aspirations

The publication of Oxford ethologist Richard Dawkins' *The Selfish Gene* in 1976 can be seen as a landmark designating the rise of the molecular imaginary. Dawkins' unfaltering focus on the gene as the evolutionary subject was instrumental in introducing both specialist and non-specialist audiences to the idea of evolution as a narrative about genes.[5] In *The Selfish Gene*, Dawkins challenges theoretical models that understand selection as involving groups or species and instead proposes the gene as the unit of selection. Seen from the "gene's-eye view" (Dawkins [1976] 1999: x), Dawkins insists, organisms are mere "survival machines" that genes build and manipulate in order to guarantee their own survival (Dawkins [1976] 1999: vii). It is, then, genes rather than organisms or species that qualify as narrative actors. In Dawkins' molecular world, organisms exist for the gene, which "leaps from body to body down the generations, manipulating body after body in its own way and for its own ends, abandoning a succession of mortal bodies before they sink in senility and death" (Dawkins [1976] 1999: 34). Literary scholar N. Katherine Hayles (2001) argues that this shift of agency from the organism to the gene is symptomatic of the fracturing of the liberal humanist subject in the postmodern age. For Hayles, "*The Selfish Gene* is underwritten by two imperatives: preserving the autonomous agency characteristic of the liberal subject, and re-locating it in the non-conscious modular units of the genes" (Hayles 2001: 150). From this point of view, the introduction of genetic agency operates as a defense against a chaotic world.

The Selfish Gene was published in the middle of the raging sociobiology controversy triggered by the 1975 publication of Wilson's *Sociobiology*. Dawkins himself saw his popular debut both as an original contribution to science and as a popularization of scientific ideas, envisioning a readership consisting of "the layman," "the expert," and "the student" (Dawkins [1976] 1999: vii–viii). While the initial reception in Dawkins' home country, the United Kingdom, was generally favorable, *The Selfish Gene* was soon immersed in the heated debate over sociobiology. Such politicization was inevitable: never shy of polemics, Dawkins transgressed culturally sanctioned boundaries by insisting that "[t]he gene is the basic unit of selfishness" (Dawkins [1976] 1999: 36), and that we should view an organism such as "a mother as a machine programmed to do everything in its power to propagate copies of the genes which ride inside it" (Dawkins [1976] 1999: 123). In her overview of the sociobiology controversy, Ullica Segerstråle (2000: 69–78) notes that the criticism emerging in the US was primarily political, casting both Dawkins and Wilson as reactionary male chauvinists, while the British criticism was primarily moral and concerned with Dawkins' unorthodox use of such concepts as free will and selfishness. What added fuel to the dispute was the fact that Dawkins' portrayal of the gene as selfish was immediately appropriated and endorsed by other sociobiologists (van Dijck 1998: 92).

Although in many ways the polar opposite of the macro-level evolutionary narratives examined in the previous chapter, Dawkins' molecular evolutionary narrative nevertheless shares a crucial similarity with these scientific epics. Like Wilson's *On Human Nature* and the Epic of Evolution and philosophical naturalist texts, *The Selfish Gene* positions itself as providing the new foundational narrative that will change our understanding of life. Like these other texts, Dawkins' book adopts a semi-religious tone. As Dorothy Nelkin and Susan Lindee insightfully observe:

> Dawkins may seem materialist and antireligious, but his extreme reductionism, in which the DNA appears as immortal and the individual body as ultimately irrelevant, is in many ways a theological narrative, resembling the belief that the things of this world (the body) do not matter, while the soul (DNA) lasts forever.
>
> (Nelkin and Lindee [1996] 2004: 53)

In *The Selfish Gene*, the first self-replicating entities from which life emerged are portrayed as having "created us, body and mind; and their preservation is the ultimate rationale for our existence" (Dawkins [1976] 1999: 20). Consistent with this religious framework, the true creators, the first replicators, and their contemporary descendants, the genes, are granted divine omnipotence, as suggested by the narrator's mock-biblical statement that "DNA works in mysterious ways" (Dawkins [1976] 1999: 21).

As with most of the evolutionary narratives in Chapter 2, the narrative logic that organizes Dawkins' account of the divine rule of the genes is based on what José van Dijck calls "the intrinsic after-the-fact tautology" that insists that "those things that survive do so because they are the fittest, but the way we define 'fit' is through survival" (van Dijck 1998: 94–95). *The Selfish Gene* differs from most earlier and contemporary evolutionary accounts, however, in how it rewrites this retrospective logic as one of fundamental, all-permeative selfishness. For Dawkins, opportunistic, fitness-enhancing gene-level selfishness is the driving force of evolutionary change and the ensuing narrative. This revision of the evolutionary logic itself returns to the competitive landscape of Herbert Spencer's (1864) infamous "survival of the fittest." At the same time, Dawkins' rewriting of self-centered individualism as a gene-level imperative stands in striking contrast to Darwin, who maintains that "[s]elfish and contentious people will not cohere, and without coherence nothing can be effected" (Darwin [1879] 2004: 155). Darwin and Dawkins situate selfishness at different narrative levels: that of the group versus that of the gene. They also give it a different narrative function. Whereas Darwin imagines selfishness as non-adaptive and thus as a narrative impediment that puts the continuity of the evolutionary narrative at risk, Dawkins understands selfishness as the engine that pushes the evolutionary narrative through change and toward futurity.

Dawkins' way of investing genes with agency necessitates that genes are understood as individual, countable, self-sustaining entities. Indeed, Hayles argues that "Dawkins's gene is the ultimate individual, the triumphant product of that brand of Anglo-American ideology that ignores the complexities of social and economic contexts and declares success or failure to be solely the result of individual initiative" (Hayles 2001: 150). Dawkins himself defends his treatment of the gene as a metaphoric individual by asserting that such textual strategy helps demonstrate complicated theoretical ideas to non-specialist audiences. At the same time, he sees experimentation with metaphor as a fertile ground for theoretical innovation:

> I prefer not to make a clear separation between science and its "popularization". Expounding ideas that have hitherto appeared only in the technical literature is a difficult art. It requires insightful new twists of language and revealing metaphors. If you push novelty of language and metaphor far enough, you can end up with a new way of seeing. And a new way of seeing, as I have just argued, can in its own right make an original contribution to science.
>
> (Dawkins [1976] 1999: xi)

Ullica Segerstråle defends Dawkins' use of anthropomorphism on precisely such grounds: Dawkins, she maintains, evokes molecular anthropomorphism in order to demonstrate how a hypothetical "*model* of a *strategy*" functions in the evolutionary dynamic of change (Segerstråle 2000: 71). We should not, she insists, think that Dawkins really believes in or wishes to advocate a view of genes as molecular homunculi. From this point of view, Dawkins' portrayal of the early replicators and the contemporary genes as masculine-coded "Chicago gangsters" (Dawkins [1976] 1999: 2), "founding fathers" (Dawkins [1976] 1999: 18), and "policy-makers" (Dawkins [1976] 1999: 60) who direct how survival machines "gamble" (Dawkins [1976] 1999: 55) in "the casino of evolution" (Dawkins [1976] 1999: 55) and speculate on "the best investment policy for a parent" (Dawkins [1976] 1999: 127) is mere rhetorical icing that can be added or removed without altering the cake.

While Segerstråle is correct in emphasizing that Dawkins is first and foremost concerned with outlining the logic of evolution, she overlooks the fact that metaphors are never devoid of cultural baggage and therefore always have consequences. Roof captures this textual dynamic when she observes that

> metaphor and analogy always import connotations and suggestions that cannot be recontained. Like the springy prank snake that leaps out of the

box, it is difficult to tuck such an excess of meaning back in place. But the problem, even for a thoughtful author like Dawkins, is that the metaphors we "lapse" into are themselves already predefined and overdetermined. The metaphors we are likely to employ are models that we have chosen in part because of the ideas and values they suggest.

(Roof 2007: 121)

Intentional or not, the "springy prank snake" of genetic anthropomorphism evokes a number of cultural assumptions about agency, responsibility, and individualism. Among these assumptions are, for example, culturally normative ideas about gender, which typically associate the attributes of individualism and initiative characteristic of the Dawkinsian gene with normative masculinity, an association reinforced by Dawkins' explicit evocation of "Chicago gangsters" and "founding fathers" above. As a result, the way in which Dawkins invokes "the fundamental nature of maleness and femaleness" while portraying molecular entities as the primary actors supports the idea that gender differences are located in the genome and, conversely, that the genome is the product of those differences (Dawkins [1976] 1999: 140).

This process of naturalization is given further force by the representation of the genes as "the immortals" (Dawkins [1976] 1999: 34). Dawkins' genes not only "manipulate" and "abandon" (Dawkins [1976] 1999: 34) like their human counterparts but also seek to guarantee eternal life. Genes, then, embody a fundamental paradox by appearing as both human-like and divine, with the result that the particular gendered humanness they invoke appears as a foundational truth. Intriguingly, Dawkins' text extends this mixture of familiar anthropomorphism and divine immortality beyond the gene. Although the gene is the true locus of subjectivity in Dawkins' evolutionary narrative, the blurriness of the different levels of biological organization and the consequent association of genes with other minuscule entities inside the body turn such entities as gametes or chromosomes into distinct (if somewhat weak) evolutionary subjects. This subtle redistribution of agency also involves the transfer of assumptions of longevity onto units with a relatively short life-span. While *The Selfish Gene* insists that "individuals and groups are like clouds in the sky or dust-storms in the desert" (Dawkins [1976] 1999: 34) and that "[c]hromosomes too are shuffled into oblivion, like hands of cards soon after they are dealt" (Dawkins [1976] 1999: 35), the text nevertheless invests a number of minute entities with a mind engaged in an opportunistic search for genetic posterity. Gametes, for example, appear as plotting for their own survival through generations (Dawkins [1976] 1999: 142–143), and the brain is portrayed as an "executive" in charge of carrying out such plans (Dawkins [1976] 1999: 60). Crucially, this extension of immortality to short-lived bodily entities both confirms the logic of interior agency that robs humans of control over their own bodies and poses a risk for the coherence of that very same anthropomorphic logic. In this sense, Dawkins' genetic anthropomorphism constitutes both the strength and the weakness of his evolutionary narrative.

Dawkins' extreme genetic anthropomorphism is further undermined by the conceptual vagueness of the gene as a measurable and separable entity. In *The Selfish Gene,* the same anthropomorphic entity that opportunistically manipulates human bodies is also explicitly "defined as a piece of chromosome which is sufficiently short for it to last, potentially, for *long enough* for it to function as a significant unit of natural selection" (Dawkins [1976] 1999: 35–36). The gene is also fundamentally inseparable from other genes as it "will overlap with other genetic units," "include smaller units," and "form part of larger units" (Dawkins [1976] 1999: 29). Rather than a mere cosmetic failing, Hayles argues, this contradictory representation of the gene as an independent actor and as a convenient scientific shorthand is what enables Dawkins' evolutionary narrative in the first place:

> If the selfish gene were only a metaphor that could be discarded at will, the motive force driving the narrative would collapse. Without this narrative, there are only shifts in populations that can be statistically measured but not causally explained. The entire argument, then, depends on the narrative in which the selfish gene, far from being a mere rhetorical flourish, is the constitutive actor.
>
> (Hayles 2001: 148)[6]

Dawkins' molecular evolutionary narrative, then, is fundamentally paradoxical, relying on two mutually contradictory conceptual frameworks. Neither framework could be excluded: there is no full-fledged gene's eye view without at least a degree of competitive, individualistic anthropomorphism, which in turn relies on the scientific understanding of evolution as a matter of calculations of probability and game theoretical management of gains and losses. Both conceptual frameworks have a further persuasive function: while scientific discourse provides the molecular evolutionary narrative with its epistemic armor, the evocation of popular discourse guarantees a wider cultural accessibility.

Dawkins' evolutionary narrative, then, is no less devoid of structural contradictions and discursive fractures than the evolutionary narratives explored in the previous chapters. Yet *The Selfish Gene* achieves a sense of coherence through its focus on the molecular level. Portrayed as human-like yet symbolically immortal, Dawkins' selfish genes are able to function as reassuringly familiar narrative agents that keep the chaotic processes and eon-spanning scope of evolution seemingly in control. That is, they operate as narrative anchors that produce a sense of direction through their own stability. In so doing, they extend and develop the coherence-building narrative role of immutable gender that we saw in the previous chapter. Furthermore, Dawkins' evocation of an interior, minuscule universe discernible only (if at all) through cutting-edge visual technologies also functions as a claim to epistemic privilege. Even if the link between Dawkins' selfish gene and visual technology is one of association rather than actuality, the popular equation of the iconic double helix with the high-tech and the cutting edge gives the molecular evolutionary narrative cultural authority associated with technological advance. In this sense, the text's emphasis on the gene also operates as a rhetorical claim of having captured the fundamental and irreducible in evolution.

The gender politics of Dawkins' molecular narrative echo Wilson's *On Human Nature*. Like Wilson, Dawkins posits a fundamental gender difference as the driving force of evolutionary change. Building on the theory of eggs as energy consuming and sperm as cheap and, consequently, of males as promiscuous and females as monogamous—a theoretical framing that will be discussed in more detail in the next chapter—Dawkins asserts that "[t]he female sex is exploited, and the fundamental evolutionary basis for the exploitation is the fact that eggs are larger than sperms" (Dawkins [1976] 1999: 146). Indeed, Dawkins argues, "it is possible to interpret all the other differences between the sexes as stemming from" this gametic asymmetry (Dawkins [1976] 1999: 141). While such a look into the cellular strategies of the gendered gametes sits rather poorly within the macro-level evolutionary trajectory that Wilson examines, it fits perfectly within the microscopic framework of *The Selfish Gene*. Dawkins' focus on the primary and the irreducible gives his portrayal of gender differences a sense of incontestable foundationality, no matter how short lived individual gametes may be compared to the scope of the evolutionary narrative. The only thing that precedes Dawkins' molecularized gender is the principle of selfishness, which, in a wonderfully circular rhetorical twist, is established as a foundational truth through its molecularized gendered manifestations.

The inner life of Adam

Dawkins' tale of genetic agency may seem like an extreme case of molecular anthropomorphism. While his extensive use of anthropomorphic imagery and culturally loaded metaphor to describe molecular processes was indeed exceptional at the time, the molecular evolutionary narrative *The Selfish Gene* introduced has become a standard textual strategy in today's popular scientific discourse. We shall turn now to two popular science books, Bryan Sykes' *Adam's Curse* ([2003]: 2004) and Natalie Angier's *Woman* ([1999] 2000), which build on the narrative foundation laid by Dawkins. Sykes' and Angier's texts make an interesting pair for comparison since they both explore the role of gender in evolution while focusing, respectively, on maleness and femaleness.

The most important cultural context for Sykes' and Angier's texts is evolutionary psychology. Evolutionary psychology combines adaptationist theory and cognitive psychology in its view of human cognition as based on distinct "modules" selected by evolution. Accordingly, human behavior is conceived as a set of evolutionary adaptations. The field arose in the early 1990s through the work of such researchers as anthropologist John Tooby, psychologist Leda Cosmides, and psychologist David Buss. One of the early key texts of evolutionary psychology was *The Adapted Mind: Evolutionary Psychology and the Generation of Culture* (1992), edited by Cosmides, Tooby, and anthropologist

Jerome Barkow, and published by Oxford University Press. Recent advocates of evolutionary psychology have included, among others, Harvard cognitive psychologist and linguist Steven Pinker, the author of bestselling and widely praised books such as *How the Mind Works* (1997) and *The Blank Slate: The Modern Denial of Human Nature* (2002). Although the scientific status and ideological commitments of evolutionary psychology are still widely debated, one thing is beyond doubt: evolutionary psychologists have been extraordinarily successful in appealing to non-scientific audiences, as suggested by the prominence of evolutionary arguments of gender and sexuality across the media.

While evolutionary psychologists often represent their discipline as clearly distinct from the sociobiology of the 1970s, many observers have emphasized continuities between sociobiologists' and evolutionary psychologists' approaches to human nature. Philosopher Val Dusek (1999), for example, refers to evolutionary psychology as "sociobiology sanitized" and queer scholar Roger N. Lancaster (2006: 110) calls it "sociobiology lite." Both critics argue that while evolutionary psychology has toned down some of the explicitly racist and sexist claims made by sociobiologists and dressed its central premise in a more palatable form, its view of mental and psychological traits as largely fixed in what is referred to as the Environment of Evolutionary Adaptedness (EEA) reflects and reinforces the central arguments of sociobiology. Like sociobiological narratives, evolutionary psychological accounts of human descent are organized according to an adaptationist logic that considers behaviors and preferences in terms of the evolutionary function they may once have played. Such claims of innate proclivities have often sought support in the discoveries of the Human Genome Project and related genetics research, as the emergence of the two enterprises coincides historically.

Sykes' *Adam's Curse* and Angier's *Woman* advocate radically different views on evolutionary psychology, as Sykes endorses and Angier critiques evolutionary psychological claims of gender differences. Published in 2003, *Adam's Curse: A Future without Men* is both typical and atypical of its genre. While Sykes' book shares with most popular science books a heavy reliance on narrative, it departs from the majority of texts in the extent to which it foregrounds its own narrative aspirations. *Adam's Curse* is simultaneously a narrative about the successes of molecular biology as a discipline, a first-person account of a particular scientist's struggles and triumphs (with emphasis on the latter), and a narrative about nature, all typical principles of narrative organization in popular science, as many scholars have observed (Myers 1990; Curtis 1994; Mellor 2003). A professor of genetics at Oxford, Sykes belongs to the growing group of practicing research scientists who are involved in explaining science from within the profession. His engagement with popularization, however, goes further than with most scientists. Sykes is a well-known figure in the British popular media as well as an entrepreneur whose company Oxford Ancestors sells genetic ancient ancestry tests to consumers.

Adam's Curse is a sociobiological investigation of the history and function of the human Y-chromosome. Through an exploration of the genealogy of his own Y-chromosome (the "Sykes Y-chromosome") and the Y-chromosomes of two historical figures—Genghis Khan and the Scottish hero Somhairle Mor—the narrator speculates on the connection between the predominance of particular Y-chromosomes in today's male population and the stereotypically masculine characteristics (bravery, aggression, competiveness, leadership skills) that the early carriers of those chromosomes supposedly exhibited. At the same time, the book sketches an apocalyptic vision of the contemporary Y-chromosomes as suffering from the side effects of the greed, lust, and aggression promoted by their molecular ancestors, and thus of the whole of humanity as facing the extinction of men in some 125,000 years. This view is based on the premise that unlike other chromosomes that come in pairs (one from each parent), the Y-chromosome is a "lonely chromosome" (Sykes [2003] 2004: 19), as men carry only one Y-chromosome. During *meiosis* (the splitting of the two sets of chromosomes in order to produce gametes with a single set of chromosomes), chromosome pairs exchange genes, a process that enables some repair of harmful mutations. The lonely Y-chromosome, however, has no such cure available, as it exchanges only a small number of genes with the X-chromosome. The Y-chromosome has therefore been subject to the slow accumulation of malignant mutations.[7] Building on Dawkins' portrayal of the most selfish genes as evolutionary survivors, Sykes suggests that the exceptionally widespread Y-chromosomes are so common because they have accumulated selfish mutations that translate into self-centered (and ultimately destructive) behavior.

The evolutionary narrative in *Adam's Curse* involves two narrative levels. The first narrative is that of macro-evolution, which is concerned with the survival and possible extinction of men, if not quite as a species, then as a species-like entity. Represented by the metaphorical Adam of the book's title, the symbol for the historical community of all past and future men, this overarching narrative level is concerned with the eon-spanning evolutionary trajectory, and, in particular, its potential endpoint: extinction. In a true socio-biological manner, this evolutionary trajectory is portrayed as predestined, written on the Y-chromosomes that all men carry. The second narrative is that of the molecular microcosm inhabited by the Y-chromosomes that the historical heroes and men living today share. Whereas *The Selfish Gene* imagines evolution as a narrative of genetic agency, *Adam's Curse* rewrites evolution as an epic history of equally human-like Y-chromosomes. Like Dawkins' genes, Sykes' Y-chromosomes are represented as unusually long-lived, as they are not fully subject to the shuffling of genetic material in the production of gametes. Crucially, Sykes is not alone in his anthropomorphic portrayal of male extinction as written in the Y-chromosome. For example, the *New York Times* science writer Nicholas Wade (1997) refers to Y-chromosome as "a wasteland" and a "desolate genetic terrain" and talks anthropomorphically about "Y's decision not to recombine" in an article somewhat confusingly titled "Male Chromosome Is Not a Genetic Wasteland, After All."

The molecular evolutionary narrative is established as the central narrative level already in the opening chapters, which depict the narrator's quest for his own patrilinear genealogy through an investigation of his Y-chromosome. By the end of the first chapter, the text has already identified genetics as the privileged field of inquiry that is able to provide the scientist-narrator, standing in Yorkshire by the site where his paternal ancestor had lived, with the exceptional "sensation" that "the Y-chromosome that I carry in all my cells had actually been here, in this place, in the fields beside the stream" (Sykes [2003] 2004: 18). Significantly, it is not the paternal ancestor but the Y-chromosome that is constructed as the narrative actor, as the one who "had been here." Moreover, *Adam's Curse* departs from *The Selfish Gene* in that it not only invests genetics with epistemic privilege but also portrays it as emotionally fulfilling, as providing a true sense of belonging. As a result, the molecular emerges as the level at which truth, understood as a unique combination of intellectual knowledge and subjective experience, resides.

The primacy of the molecular and minuscule is further emphasized through reference to visual technologies. *Adam's Curse* reaches a crucial climactic point when, after a series of complicated technical procedures described in detail and with admiration, the narrator looks at his own Y-chromosome under the lenses of a microscope:

> This is my Y-chromosome, the bearer of my maleness and the token passed unaltered down from a long line of fathers. This is the chromosome I have come to see. I see it in my own father, as he leads his RAF squadron in the Second World War. I see it in my grandfather, fighting in the trenches and wounded in the battle of the Somme a generation earlier.
>
> (Sykes [2003] 2004: 29)

Even though this scene takes place early in the text (the end of Chapter 2), it is arguably the most important moment in the whole 300 pages of *Adam's Curse*. First, it marks the point at which the molecular world takes precedence over other narrative levels. This is reached by turning the previously abstract and theoretical molecular microcosm into a world that is visual and thus, by association, both real and within our grasp. Second, despite its early occurrence in the text, this is also the moment when the narrator's quest for his own paternal genealogy finds its completion in the molecular vision. It is in the minute, in the colorful shapes under the high-powered microscope, that the narrator locates the truth. The same fixedness that made the Y-chromosome so vulnerable to mutation turns out to be where its uniqueness lies.

While echoing Roof's observation about the valorization of the minute in scientific inquiry, such a detailed description of the technologically assisted production of interior vistas also reinforces the common belief that technology is the key to fundamental insight. In his survey of popular science books, Jon Turney observes that popular science texts are faced with the challenge of producing "with everyday words on the page" the feeling that the science the text introduces really does what it claims to do (Turney 1999: 125). According to Turney:

The writer has to *describe* observations, experiments, reports, even demonstrations, in such a way that the reader believes in them, and believes that they signify what is claimed ... [R]eproduction of these experiences through text requires, above all, success in *persuasion*—success in persuading readers that reality is as it is described and success in persuading them that they understand it.

(Turney 1999: 125)

Sykes' evocation of visual technologies appropriates the hype surrounding biotechnologies, thereby associating knowledge with technology and truth with the minute and interior. In doing so, the text buttresses the assumed epistemic authority of genetics. Most importantly, however, Sykes' depiction of his chromosome blurs the line between science and its description. As Sykes' emotionally loaded "everyday words" stand both for the revelation of the truth hidden inside the cell and for the narrator's personal experience, his experience becomes equated with the objective and authoritative gaze of science. This becomes evident in Sykes' use of the verb *see* in the quotation above. While "seeing" is established in this scene as professional observation of a scientific experiment (seeing things through the microscope), it also comes to stand for subjective imagination, as the narrator "sees" the chromosome in his father and grandfather as well as "sees" his father and grandfather fighting in the two World Wars. This curious mixture of technology mediated vision, personal memory, and pure imagination resonates with José van Dijck's (1998: 167) observation that visual representations of sequenced genetic material are often interpreted as photographic "snapshots" of a person's genetic inheritance in popular discourse.

Once established in the text, the molecular evolutionary narrative appropriates features of what Abbott (2003) describes as the narrative levels of species and organism. "Passed unaltered" from paternal ancestors, Y-chromosomes appear as entities that are, if not quite immortal, at least remarkably long-lived. Like Dawkins' genes, they capture the temporal scope implicit in the macro-level narrative of the never-ending evolution of diverging species. At the same time, the same Y-chromosomes appropriate characteristics usually attributed to organisms. While it was the fathers who fought bravely in the war in the passage quoted above, it is the chromosomes that do the fighting once the molecular evolutionary narrative has been established as the primary narrative level. The prominence of certain Y-chromosomes in today's male population leads the narrator to speculate whether these Y-chromosomes demonstrate some unusual characteristic. Pondering the wide geographical spread of what he assumes to be the "Genghis Khan Y-chromosome," the narrator asks rhetorically: "Is the Khan chromosome's achievement down to the sexual exploits and military conquests of the Mongol emperor? Or was the Great Khan himself driven to success in war, and in bed, by the ambition of his Y-chromosome?" (Sykes [2003] 2004: 187). In Sykes' molecular universe, it is the Y-chromosomes that have become the subjects of the "little stories of love and death" that Abbott ascribes to organisms. In doing so, they bridge the structural gap that separates the abstract Adam of the species-level narrative from the historical men who live, mate, and die. As a result, Sykes' molecular view of evolution emerges as a coherent narrative that combines historical scope, even teleology, with plots of love, lust, and treachery.

The Selfish Gene demonstrated that the molecular evolutionary narrative tends to naturalize gender ideologies. *Adam's Curse* takes this process still further through its consistent and immediately recognizable use of gendered cultural vocabulary. Projected on the Y-chromosome, such stereotypically masculine attributes as the bravery and endurance demonstrated by the narrator's father and grandfather in the two World Wars become seen as rooted in molecular processes. Through the positioning of the "Khan Y-chromosome" as an anthropomorphic narrative actor, the military skills and polygynous sexuality associated with the Mongolian emperor appear as arising from male biology. Sykes' representation of maleness as opportunistic and aggressive, and opportunism and aggression as written in the Y-chromosome, reflects popular scientific discourses of maleness. For example, a *New York Times* article titled "Y Chromosome Depends on Itself to Survive" reports on a discovery that the Y-chromosome "recombines with itself" (2003). The article refers to this phenomenon as a *"narcissistic* process of salvation" that "has saved men from extinction so far" (Wade 2003; emphasis mine).

This process of naturalization extends beyond maleness. The very idea of sexual difference itself is established as a molecular-level truth through a similar two-step process. The Prologue to *Adam's Curse* depicts the genders as strictly binary:

> The fact that we humans exist in two forms is so much part of everyday life, and always has been, that we rarely pause to question why this should be. Yet, the simple distinction between male and female divides our species into two perennially polarized camps separated on either side of a great canyon from whose rim we signal to each other and struggle to hear, but which we can never cross.
>
> (Sykes [2003] 2004: 2)

Gender differences, the passage suggests, are natural, fundamental, and self-evident, a "fact" we all supposedly accept. As with stereotypical ideas of heroic masculinity, these two "perennially polarized camps" become projected on the minute narrative world. By the end of Chapter 4, the differences between the genders have become equated with molecular processes "embedded deep within our genomes" and men and women have been replaced by the dichotomously positioned Y- and X-chromosomes as narrative actors (Sykes [2003] 2004: 3). This association of the X-chromosome with femaleness and the Y-chromosome with maleness is, of course, misleading since both men and women carry X-chromosomes. In this sense, Sykes' text appropriates the popular idea of X- and Y-chromosomes as "sex chromosomes."

In *The Descent of Man*, Darwin describes gender differences as a product of evolution, especially the processes of sexual selection. According to Darwin, "[w]e may conclude that the greater size, strength, courage, pugnacity, and energy of man, in comparison with woman, were acquired during primeval times, and have subsequently been augmented, chiefly through the contests of rival males for the possession of the females" (Darwin [1879] 2004: 674). At the same time, these gender differences also function as the impetus of sexual selection. In *Adam's Curse*, this same logic becomes translated into molecular processes within the cell. Trapped on the opposite sides of the great canyon that dominates the evolutionary landscape are not the organisms but their anthropomorphic chromosomes. These chromosomes are, however, also the very agents of sexual selection. Looking for a "dancing partner" for "the final dance" (Sykes [2003] 2004: 43) and seeking to "come together for a final embrace" (Sykes [2003] 2004: 42) and "intimate exchanges" (Sykes [2003] 2004: 43), Sykes' human-like chromosomes reproduce a particular, culturally sanctioned heterosexual dynamic. This projection of the discourse of romance onto gendered chromosomes closely echoes the romantically involved egg and sperm Emily Martin (1991) identifies in textbooks on reproductive biology. At the same time, the dichotomously organized heterosexuality exhibited by Sykes' X- and Y-chromosomes becomes the very engine that keeps the evolutionary narrative going.

This picture of men and women as fundamentally unlike each other is further emphasized through a rhetorical maneuver that positions science as the privileged account of all phenomena, whether molecular or social:

> The more I dug down, the more I realized that the two sexes are caught in a dangerous genetic whirlpool, playing out in the flesh irreconcilable conflicts embedded deep within our genomes. As much by luck as judgement, my own research on DNA placed me in a unique position to observe this primal struggle. I found myself with the means to follow the different genetic histories of men and women. I could listen to the messages carried by DNA and catch the whispers of old lives passed on by generation after generation of ancestors.
>
> (Sykes [2003] 2004: 3–4)

Read in the context of the earlier passage about men and women "struggling to hear" each other from the opposite sides of the gendered canyon, this quotation invokes two closely related metaphors. On the one hand, men and women are represented as speaking different languages and thus as unable to comprehend each others' sexual syntax. This theme of a total communicative break between the genders is familiar to popular audiences especially from John Gray's 1992 bestseller *Men Are from Mars, Women Are from Venus*, insightfully analyzed by Annie Potts (2002: 48–68). On the other hand, molecular biology is portrayed as an interpreter of these two foreign tongues. Able to "listen to the messages carried by DNA and catch the whispers of old lives," science is understood as a task of deciphering, translating, and interpreting, an association that relies on the popular view of DNA as a cryptic historical document. Through this dual use of the language metaphor, *Adam's Curse* reinforces popular assumptions of the genders as fundamentally different and of science as an epistemically privileged enterprise that is able to reveal the final truth about gender relations.

All in all, *Adam's Curse* appropriates and develops narrative potential implicit but not fully articulated in *The Selfish Gene*. Translating the idea of the gene as immortal and agential into a portrayal of Y-chromosomes as long-lived and opportunistic, Sykes relies on the metonymic transfer of attributes (immortality, subjectivity) associated with one level of biological organization (the gene) to another (the chromosome). As a result, the Y-chromosomes appear as both astonishingly exceptional yet fascinatingly familiar, while the molecular universe emerges as the locus of truth and molecular biology as the privileged science capable of interpreting that truth. *Adam's Curse* also turns Dawkins' tentative use of gendered anthropomorphism into a full-fledged portrayal of the human interior as a thoroughly gendered microcosm. This revision and extention of Dawkins' molecular narrative attests to the imaginative potential it carries. Finally, Sykes follows Dawkins' representation of evolution as ambitious narrative movement toward futurity, produced, paradoxically, through the immutable gender of its molecularized actors. Yet Sykes' way of emphasizing the danger of extinction gives the molecular evolutionary narrative a sense of uncertainty that subtly undermines the finality of gender. This does not, however, open up the category of maleness for critical redefinition, as the text's anthropomorphic rhetoric equates self-assertive masculinity with molecular truth. Instead, futurity appears as a question of aggressive, ambitious men, or no men at all.

Eve's revenge

Sykes' way of rewriting popular plots about dominating men and coy women as foundational narratives of molecular-level desire and death situates *Adam's Curse* firmly within the tradition of gene-centered sociobiology and evolutionary psychology. Do evolutionary psychological claims, then, always accompany the molecular evolutionary narrative? Do texts that critique evolutionary psychology need to reject the molecular evolutionary narrative as a textual strategy? The answer to both questions is not necessarily. Natalie Angier's *Woman: An Intimate Geography*, published in 1999, sports such a critique. Angier's book portrays evolutionary psychology as "a cranky and despotic Cyclops, its single eye glaring through an overwhelmingly masculinist lens" (Angier [1999] 2000: 355). Rather than aligning her text with social constructionism, however, Angier appeals to scientific explanations in support of her argument. As a Pulitzer Prize-winning science writer for the *New York Times*, Angier also differs radically from Sykes the Oxford professor as a narrative authority.

If Sykes' *Adam's Curse* presents a narrative quest for "true" maleness, Angier's *Woman* is an appraisal of the complexity of the female body. Characterized in the introductory chapter as "a celebration of the female body—its anatomy, its chemistry, its evolution, and its laughter," *Woman* consists of 19 chapters that each speculate on the function and evolution of a given female-specific organ (such as the ovary or uterus), female-associated bodily entity (such as the egg or X-chromosome), or a gendered behavioral pattern (such as aggression) (Angier [1999] 2000: xiii). Like *Adam's Curse*, *Woman* evokes multiple levels of the evolutionary narrative, although it emphasizes them differently. As Angier's book is primarily concerned with the here and now, the narrative of species-level change, rewritten as the descent of "Woman," is less prominent than the narrative of men as a species-like entity in Sykes' text. *Woman* invokes this macro-level narrative when discussing the evolution of a particular organ or molecular entity, and thus it consists of a number of separate narrative lines that each lead to the evolution of the given organ or molecular entity. These storylines do not always form a final or unambiguous evolutionary trajectory, as Angier—unlike Sykes—introduces and discusses contradictory theories, and the coherence is often implicit. Contrary to *Adam's Curse*, *Woman* also gives considerable space to Abbott's "little stories of love and death" of individual organisms, as the text repeatedly introduces real, living women as examples of a particular biological phenomenon. This textual device echoes the journalistic practice of giving social or global phenomena a human face by highlighting individual destinies and experiences. These "case studies" have two functions: they evoke empathy in the reader and serve as strategic introductions to the scientific phenomena explored in the text.

In *Woman*, the stories of individual women operate as gateways to the interior cosmos of the female body. It is through these journalistic case studies that the female-only interior universe is established as the primary narrative site. The interior microcosm of *Woman* is more comprehensive than that in *Adam's Curse*, covering the whole interiority of the female body. Where *Adam's Curse* extends attributes of agency from genes to chromosomes, *Woman* appropriates this metonymic potential even further, representing various levels of biological organization within the body as a coherent, agential whole. The contrast between the exterior world of real women and the interior world of the body gives the latter an air of completeness, rendering invisible the structural differences among the mutually incompatible bodily entities. This effect is produced through the deployment of anthropomorphic verbs that help construct these entities as narrative subjects. We encounter, for example, chromosomes that "behave with great courtesy" (Angier [1999] 2000: 21), a cell that "makes its own decision" (Angier [1999] 2000: 24), a clitoris that "knows" (Angier [1999] 2000: 79), and a breast that "arouses" (Angier [1999] 2000: 139). These entities even communicate in a humanlike fashion, as when "a follicle, on feeling the temblor, will quicken its pace of maturation and tell the brain, Hurry up, please, it's time" (Angier [1999] 2000: 191).

As in *Adam's Curse*, the molecular narrative is introduced early. The very first chapter of *Woman* describes the extraction of eggs from a donor's ovaries in the following words:

> [Dr] Bustillo does the entire extraction procedure by watching the ultrasound screen, where the image of the ovary looms in black and white, made visible by bouncing high-frequency sound waves. Coming in on the top lefthand side of the screen is the needle. The ovary looks like a giant beehive honeycombed with dark bloated egg pockets, or follicles, each measuring two millimeters across. These are all the follicles that were matured by Derochea's [the donor] diligent nocturnal injections. The sonogram screen is full of them. Manipulating the needle-headed probe with her eyes fixed on the sonogram, Bustillo punctures every dark honeycomb and sucks all the fluid out of the follicle.
>
> (Angier [1999] 2000: 12)

The passage moves from a case study (the altruistic egg donor Derochea introduced earlier in the same chapter) to the interior world of the body, a narrative strategy that is repeated throughout the book. As in *Adam's Curse*, the micro-level narrative the scene introduces is given a sense of realness and immediacy through the evocation of visuality. Here, too, technology provides a unique means of viewing the hidden interiority of the body, a representation underlined by the detail in which the procedure is described. These similarities between Angier's description of the human egg and Sykes' description of the Y-chromosome also suggest that interior vistas produced through different forms of visual technology—in this case, sonogram and microscope, respectively—are often interpreted as one and the same in popular discourse. It is as if technology as a mediator engendered a curious loss of the sense of distance.

Once established as a central narrative level, Angier's revised version of the molecular narrative takes on characteristics from the narrative levels of organisms and species. As in *Adam's Curse*, the initial evocation of visual technologies is followed by a series of anthropomorphic moves as entities like the ovary and the X-chromosome become portrayed as if they were organisms rather than their constitutive parts. At the end of the first chapter, only a few paragraphs after the ultrasonic vision of the eggs, we are already in the anthropomorphic world of microscopic units in which "[e]ggs must plan the party" while "[s]perm only need to show up—wearing top hat and tails, of course" (Angier [1999] 2000: 17). After this, a further justification for anthropomorphism hardly seems necessary in Angier's text. From the level of the slow and continuous evolution of species, on the other hand, Angier's molecular narrative appropriates notions of longevity and even teleology. The female body appears as a biological microcosm that has its own evolutionary trajectory, and within which all entities have their own historically defined functions. *Adam's Curse* and *Woman* mirror each other in this respect: while *Adam's Curse* portrays the Y-chromosome as a long-lived evolutionary hero,

Woman projects assumptions of emerging complexity often present in the eon-spanning narrative of species-level change onto microscopic phenomena such as the delicate "ecosystem" of the vagina (Angier [1999] 2000: 51–61) or the intricate biochemical dynamics that produce changes in uterine tissues (Angier [1999] 2000: 99–119). However, the books differ in the tone in which they deliver their molecular epics. While Sykes' portrayal of the male-centered microcosm is elegiac, lamenting the decline of the previously triumphant male genealogy, Angier's microscopic female universe is simply triumphant. Where *Adam's Curse* suggests nostalgia for what is perceived as a lost age of heroism, *Woman* celebrates a golden age that is always only beginning. There is, indeed, curious stability in Angier's portrayal of the female body. While highlighting the transformative power of natural selection, *Woman* invests this push toward futurity with a contradictory sense of perpetual present.

There is another crucial difference between Angier's and Sykes' appropriations of the Dawkinsian molecular narrative. While both texts portray molecular entities as human-like actors, the kinds of popular vocabularies they deploy are radically different. Whereas Sykes' Y-chromosomes appear as ambitious and aggressive warriors, Angier's microscopic entities emerge as a sort of proto-feminist sisterhood. Paralleling the implied female readership of the text—the narrator refers to her audience as "you, my sisters" (Angier [1999] 2000: xvi)—human eggs are represented as forming a female alliance. Not only do the eggs actively and independently choose the sperm they prefer but they also work for a common female—even feminist—cause. When explaining the working of a cell elimination process called *apoptosis* that reduces the number of eggs in a woman's body during her lifetime from over six million in a 20-week old fetus to 450 or so in an adult woman, the narrator describes the eliminated eggs as "sacrificial" heroines who "commit suicide" for the sake of their better-equipped "sisters" (Angier [1999] 2000: 3). This sense of female comradeship is reinforced through the representation of the female body as an evolved system, where all parts work perfectly together. Accordingly, evolution appears as a coherent and coordinated process unlike the endless, antagonistic battles among Sykes' individualistic Y-chromosomes. Moreover, the adventures of Angier's anthropomorphic heroines echo feminist empowerment narratives, as her microscopic entities take on male-associated characteristics and appropriate them to their own ends. The uterus, for example, is portrayed as a "warrior" (Angier [1999] 2000: 119) and "muscle hero" (Angier [1999] 2000: 102) and the placenta and the mammary gland as "specialists" (Angier [1999] 2000: 160), all strategic choices that rewrite ideas of bravery, strength, and skill associated with masculinity in the popular imagination.

While both *Adam's Curse* and *Woman* depict the microscopic world as clearly gendered, Angier turns the familiar gender binary upside down. In *Woman*, the minute heroines recurrently appear as equal to or more powerful than their male-codified counterparts, as when the X-chromosome is described as playing a crucial part not only in the construction of the female body but also in "the creation of Adamically correct genitals" (Angier [1999] 2000: 33).

Similarly, the egg appears as the main player in reproduction, as the entity that "must hold the secrets of genesis" (Angier [1999] 2000: 17). By contrast, the role reserved for such a male-codified entity as sperm is that of the "Trojan horse" (Angier [1999] 2000: 112) and the "gift-bearing Greeks" (Angier [1999] 2000: 113), an analogy that suggests that as an evolutionary underdog, the sperm cannot help but rely on treachery. Testosterone, too, is described as "not a particularly active hormone" (Angier [1999] 2000: 268). This failure of male activity is further rewritten as a failure of male sexuality, as suggested by the depiction of testosterone as "much less biologically *potent* than estradiol" (Angier [1999] 2000: 268; emphasis mine). In striking contrast to the fearsome and lustful (if soon infertile) Y-chromosomes that rule Sykes' molecular cosmos, Angier's male-codified entities appear as unreliable, fickle, and impotent compared to their powerful female-codified counterparts.

More important than these differences in gendered imagery, however, is the similarity of the effect that anthropomorphism produces. Despite its feminist agenda, *Woman*, too, risks naturalizing cultural imagery by projecting it onto the minute interior world. Like Sykes' molecular heroism and competitiveness, the proto-feminist spirit of Angier's microscopic sisterhood appears as arising from natural laws about sexual relations. The suicidal eggs and the heroic uterus described above, for example, seem to embody a strictly gendered set of behaviors that becomes anchored in the realm of nature. As a result, feminist solidarity and bravery cease to be political choices and emerge as essential characteristics of female biology and, by extension, real women. However, this process of naturalization is in constant tension with Angier's explicit critique of genetic determinism. The tension becomes evident, for example, when the narrator explains hormonal gender differences as a question of degree rather than kind while portraying estrogen as the femininely gendered "Lady Estradiol" (Angier [1999] 2000: 223) and "our steroid heroine" (Angier [1999] 2000: 206) and testosterone as a true macho with a "cocky little eye" (Angier [1999] 2000: 267).

It should be emphasized that Angier's treatment of scientific theories is generally complex and her use of gendered anthropomorphism often ironic and witty. This ironic tone is present, for example, when the narrator describes theories of female orgasm as "good clean dirty cortical fun, so much more amusing than being adjured, as we were in the 1970s, to get a mirror and inspect our genitals for ourselves" (Angier [1999] 2000: 71). She continues: "Some researchers have argued, in print, that the female climax may be so unnecessary as to be on its way out. One unlucky lurch of the evolutionary wheel, and those fibers may fire no more" (Angier [1999] 2000: 71). Yet the way in which Angier's molecular anthropomorphism arises from narrative structure produces a narrative momentum that undermines the theoretical plurality and ambiguity of tone. This tension between naturalization and denaturalization constitutes a crucial fracture in the evolutionary narrative that *Woman* produces. While aware of the complexity of gender and sexuality, Angier's text is not really that different from Sykes' *Adam's Curse* in terms of the effect that the molecular evolutionary narrative generates. Sykes' and Angier's evolutionary narratives are ultimately variants of the circular narrative logic of Wilson's *On Human Nature*. While operating at different narrative levels and experimenting with different textual strategies, Wilson's, Dawkins', Sykes', and Angier's texts all rely on a dynamic that understands the present as both the ultimate product and the very foundation of the unfolding evolutionary narrative. This circularity is reinforced through the tension between the stability of evolutionary actors and the unceasing movement of evolutionary processes. As Wilson's, Dawkins', Sykes', and Angier's texts all demonstrate, this seemingly contradictory juxtaposition is able to engender culturally resonant accounts that naturalize contested ideas of gender, sexuality, and human nature.

Deviant dictators

These affinities between *Adam's Curse* and *Woman* demonstrate that the Dawkinsian molecular narrative is remarkably flexible as a narrative structure. Such versatility suggests that the molecular evolutionary narrative has become a standard device in popular science. It also attests to the prominence of molecular imagery in culture at large. The rest of this chapter provides a tentative examination of the extent to which the molecular evolutionary narrative has been integrated in the cultural imagination beyond the genres of popular science. As with the macro-level scientific epic discussed in Chapter 2, this issue is explored through a close reading of a literary work. Simon Mawer's novel *Mendel's Dwarf* provides a detailed examination of the history and future of modern genetics while exploring the hopes and anxieties it raises today. Its appropriation of gendered anthropomorphism and invocation of claims of foundationality point to yet another twist in the molecular evolutionary narrative. In particular, *Mendel's Dwarf* demonstrates how the molecular narrative becomes the site of a highly charged discursive encounter between the practices of professional science and the cultural expectations about their consequences.

Published in 1997, *Mendel's Dwarf* appeared in the middle of the hype surrounding the Human Genome Project. Put briefly, *Mendel's Dwarf* explores ethical and philosophical issues raised by molecular genetics through two parallel plots. The primary narrative depicts the story of the narrator-protagonist Dr Benedict Lambert, a renowned geneticist and a great-great-great-nephew of Gregor Mendel, who embarks on a search for the gene for achondroplasia (a form of dwarfism). Ben himself is achondroplastic. This narrative is juxtaposed by an account of Gregor Mendel's escape from the hardships of peasant life to the monastery in Brno, his experiments on inheritance in the garden pea in the monastery, and his subsequent frustration when faced with the scientific establishment's dismissal of his results. Ben Lambert is the sardonic narrator of both of these narratives, a textual strategy that suggests a likeness between the early twentieth-century eugenic movement that evoked Mendel's discoveries and the questions of reproductive rights and genetic engineering made potent by the new genetics of the late twentieth century. This association of what the novel portrays as the "old" and "new" eugenics gives additional resonance to the ethical questions the novel's main storyline raises. Against his and society's expectations, Ben Lambert engages in an intimate relationship with Jean Piercey, librarian and wife of abusive and egoistic Hugo Miller, and eventually agrees to Jean's pleas for having her eggs secretly fertilized by those of his sperm that the brand new prenatal test for achondroplasia proves healthy. The novel ends with Jean's labor-induced coma, Ben's vengeful claim for paternity, and the murder of their child by the bitter and jealous Hugo, leaving unresolved the complex questions about reproductive rights, prenatal screening, selective abortion, and genetic engineering it has raised.

As important as these ethical and philosophical concerns are, I focus here on the role of the Dawkinsian molecular narrative in Mawer's novel. Rather than dismissing the question of ethics, however, such an analysis may help shed light on the discursive dynamic within which ethical questions about agency and free will arise in the first place. The molecular evolutionary narrative is present in *Mendel's Dwarf* through the emphasis placed on a particular gene's (the mutation that causes achondroplasia) control over the destiny of an organism (Ben Lambert). *Mendel's Dwarf* distances this molecular evolutionary narrative from Darwin's species-level account of natural selection by emphasizing that Mendel, unlike Darwin, understood the fundamental truth about the laws of inheritance. At the same time, Mawer's novel differs from Dawkins', Sykes', and Angier's versions of molecular agency in that it imagines the molecular from the point of view of the Dawkinsian survival machine, the organism. In *Mendel's Dwarf*, human experience appears as a captive of the genome. While this is most evident in the narrator-protagonist's portrayal of his life as determined by "the malign hand of the mutation" (Mawer 1998: 11), the ultimate high-handed dictator, he asserts that it is only by good fortune that most of us are born healthy, as we "are all victims of whatever selection of genes is doled out at that absurd and apparently insignificant moment when a wriggling sperm shoulders aside its rivals and penetrates an egg" (Mawer 1998: 22).

Mendel's Dwarf follows Dawkins' portrayal of the gene as the true narrative agent but invests this logic of predestination with a sense of coincidence. From the point of view of the organism, the passing of genes to subsequent generations seems terrifyingly chaotic, as the presence of one set of genes rather than another in a given body is ultimately dictated by the "random, destructive, but also occasionally creative Lady Luck" (Mawer 1998: 173). The significance of chance is further emphasized by the fact that the gene responsible for Ben's achondroplasia is a spontaneous mutation merely a generation ago rather than a stretch of DNA passed through centuries. This representation of humans as helpless victims resonates strongly with cultural anxieties about the invisible defects that we may carry within our genomes. These anxieties underlie, for example, the rapidly growing field of genetic tests, which are increasingly available to consumers not only through medical practitioners, but also through online genetics companies. Such tests promise to identify genetic susceptibility to certain diseases or conditions, a promise that is easily turned into a burden, as scholars studying the ethical and legal aspects of genetic testing have pointed out (Cox and McKellin 1999; Taub *et al.* 2004; Majdik 2009). Anxieties about the hidden threat of genetically controlled diseases and disabilities also colors popular cultural narratives. The Fox network's medical drama series *House*, for example, sports a troubled central character, Dr Remy "Thirteen" Hadley, whose mother has died of Huntington's disease. Thirteen eventually decides to test for the gene for Huntington's, and after the positive result tries to cope with the fact that her every cell carries an invisible genetic defect that will kill her.

As in *The Selfish Gene*, *Adam's Curse*, and *Woman*, molecular narrative agency in *Mendel's Dwarf* is not restricted to genes but is also projected onto other minuscule entities. For example, the narrator describes his father's exposure to radiation in military service as if the gamma rays were human-like agents:

> [W]hile the light was filtered by the dark glass of the goggles, the gamma rays, subtle and unseen, wafted freely through cloth and flesh and bone. In the course of their passage, did they touch with malign and featherlike hands the dividing cells buried deep within my father's testes?
>
> (Mawer 1998: 13)

Similarly, viruses (Mawer 1998: 80) and bacteria (Mawer 1998: 157) appear as invisible and mischievous agents. Gametes, on the other hand, embody the culturally sanctioned heterosexual dynamic identified by Martin: the young Ben peers through the microscope at "[o]ne million million spermatozoa" who all embody "[s]mall exclamations of blind and culpable intent" and asserts that "[t]he moment of true penetration is when the lucky sperm, the poor Noah, nudges against the ovum and explodes its capsule of digestive enzymes" (Mawer 1998: 28). As the site of the "true penetration," the hidden interior microcosm emerges as the location of true narrative agency, separated from the human world by the failure of the human eye and the necessity of mediating technology.

As in *The Selfish Gene*, *Adam's Curse*, and *Woman*, this microcosm appears as foundational through appropriation of religious imagery. Echoing the vocabulary of genetic "scripture" we encountered in Matt Ridley's *National Geographic* article in Chapter 1, Ben Lambert asserts that whereas "[o]nce upon a time the mystery was enshrined in the tabernacle on the altar, in a sliver of wafer. Now it lies, stripped open for mankind to read, in a poly-acrimide denaturing gel" (Mawer 1998: 133). Holding the key to the "secret of life" (Mawer 1998: 133), the human genome is "some thousand times as big as the entire Bible" (Mawer 1998: 134), while geneticists are "exegetes" who "have read the words of the scroll of life" (Mawer 1998: 197). Within this framework, science in general and genetics in particular replaces religion as the foundational narrative of human condition. As in the other texts, this sense of foundationality is further reinforced through the simultaneous emphasis on the primacy of the visual and the difficulty of seeing. Early in the text, Mendel is described as a man "whose vision goes beyond what we can perceive with our eyes and touch with our hands ... Mendel looked through the surface of things deep into the fabric of nature, and he saw the atoms of inheritance" (Mawer 1998: 4). Truth, this suggests, is to be found in the irreducible particles of life, the ultimate life-giving components that embody the holy grail of the history of Western science.

Mendel is able to correctly imagine these fundamental particles only because of his exceptional talent. These particles are, however, beyond the reach of the human eye, as the narrator-protagonist points out: "You see nothing. It takes place in a miniature test tube" (Mawer 1998: 157). Apart from the line between insight and failure to imagine, then, the text draws a further line between the theoretical understanding achieved by Mendel and the visual recognition attained through scientific technology. This imagined epistemic boundary becomes manifest when the narrator ponders how Mendel might have responded to "the fact that we can actually read the messages enshrined in the hereditary particles whose existence he could interfere only from watching the way they behave" (Mawer 1998: 94). This portrayal grants visual technology the status of an epistemically privileged enterprise, positioning it as witness to molecular truths. Furthermore, *Mendel's Dwarf* resembles *Adam's Curse* in its understanding of genetic genealogy as providing not only intellectual satisfaction but also a unique emotional experience. Like Sykes' narrator in Yorkshire, Ben Lambert reports on his feelings while standing by his great-great-great-uncle's monastery in Brno: "Oh yes, I felt something as I stood looking across the lawns: something stirring in the bowels as well as in the brain, something that evades the grasp of words" (Mawer 1998: 10). Even though the scene is not imagined in terms of anthropomorphic genes or chromosomes, it posits genetics as the true connector through a dual association: the sense of connection Ben experiences links two practitioners of genetics on a historical continuum as well as binds distant relatives through the narrative of shared inheritance.

Despite the novel's ethical and philosophical concerns and the narrator-protagonist's insistence that genes "don't do anything else" but "code for protein" (234), these molecular protagonists take on human characteristics. As in the other texts, anthropomorphism risks naturalizing the cultural vocabulary through which the molecular entities are produced. This effect is reinforced by the novel's invocation of a detective plot, a narrative strategy that science studies scholar Ron Curtis (1994) considers as a standard device in popular science. Like the exotic viruses and mysterious diseases that the scientists in Curtis' popular science magazines patiently seek to identify, the search for a particular mutation in the genome is depicted as a form of unrelenting detective work in *Mendel's Dwarf*. Ben Lambert's genetic mutation is "the suspect" (Mawer 1998: 135), the criminal offender whose acts are marked by inherent dishonesty, as "even DNA, that most innocent of molecules, lies" (Mawer 1998: 177). Scientific inquiry, by contrast, is the industrious "flatfooted policeman" who embarks on "a door-to-door search" in a particular "district" of the genome (Mawer 1998: 134), "looking for possible associates of the unknown man" and finally "closing in" (Mawer 1998: 135). This imagery of human victims at the mercy of dangerous genetic criminals and in need of the help of the scientist-detective renders molecular agency implicitly masculine while glorifying molecular genetics as the heroic, clever, and patient solver of the mysteries of life. Through the figure of "the flatfooted policeman," it also reinforces the old idea of science—and by implication epistemic authority—as symbolically masculine.

Notwithstanding the novel's complex take on ethical questions, then, the molecular narrative in *Mendel's Dwarf* is strikingly similar to the molecular narratives in the other texts explored above. Mawer's depiction of humans as passive victims, the molecular as the true site of agency, and technology as the location of epistemic privilege suggests that Dawkins' gene-centered perspective to life has become so common it has lost its sense of defamiliarizing distance. At the same time, the way in which *Mendel's Dwarf* replaces the Dawkinsian narrative of the survival of the fittest gene with the unpredictable appearance of a spontaneous mutation sheds light on the organizing logic of the molecular evolutionary narrative. Rather than maximizing reproductive success, the gene for achondroplasia reduces what evolutionary biologists call reproductive fitness, as it is only in secrecy that Ben gets to pass on his genes. Crucially, the gene for achondroplasia is not a recessive gene that may travel undetected through several generations but a spontaneously mutating dominant that, if passed, will always be expressed. As a spontaneous mutation, Ben's gene falls outside the eon-spanning existence of Dawkins' immortal replicators, thereby standing in curious contrast to the image of the biblical genome as the ultimate archive of human evolution.

This puts the very act of reproduction in a new light. While *Mendel's Dwarf* echoes the familiar view of reproduction both as the engine of evolutionary change and as the end-product of historical processes, it also insists that reproduction is always risky. As a Dawkinsian survival machine, Ben's body has failed the ambitions of the thousands of his healthy and presumably self-interested genes. Ben's story, then, suggests that inherent in every act of reproduction is the possibility of things going wrong. Such a fracture in the line of inheritance may provide a surprise both for the individual organisms and their imagined dictators, the immortality-seeking genes. Most importantly, however, the very existence of Ben's mutation brings significant fragility to the seemingly uncontestable logic of evolutionary movement by suggesting that narrative continuity is never fully secured. This renders the concept of transformation increasingly suspect, as a slight molecular change may have dramatic consequences on the narrative levels of both genes and organisms. We return to this issue in Chapter 5.

Molecular mythologies

This chapter has traced the emergence of the molecular evolutionary narrative in the 1970s and followed some of the narrative mutations it has undergone after the introduction of the Human Genome Project in 1990. This analysis has shown that the molecular narrative brings narrative coherence to the Darwinian evolutionary narrative previously haunted by the disconnection between organisms and species. In particular, it has produced a seemingly unified narrative agent by casting minuscule bodily entities as strictly gendered evolutionary protagonists. While Dawkins' pioneering *The Selfish Gene* focused on genes, subsequent versions of the molecular narrative have increasingly extended agency to such microscopic entities as chromosomes, hormones, gametes, and viruses. Apart from locating narrative action at a single narrative level, these new evolutionary agents differ from organisms, species, and populations in that they embody characteristics of both the slow macro-level evolution and the safe familiarity of the "little stories of love and death" of individual organisms.

The similarities among Dawkins', Sykes', Angier's, and Mawer's texts demonstrate that the molecular evolutionary narrative is not tied to a particular genre of popular science or a particular strand of evolutionary theory. The texts differ clearly in their argument, their take on sociobiology and evolutionary psychology, and the status of the author as a scientific authority, and their versions of molecular anthropomorphism appropriate different cultural vocabularies. Despite these differences, all the texts project anthropomorphic attributes onto the minute entities inside the body, thereby rendering those attributes universal biological truths. By employing notions of longevity from the narrative level of the slow evolution of species, the texts give that very anthropomorphism an air of foundationality and evolutionary necessity. This is the case even with such a considerably multivalent text as Angier's *Woman*, or a complex and highly ironic account as Mawer's *Mendel's Dwarf.* The molecular evolutionary narrative, then, is not tied to a specific view of biology or gender, but rather relies on a particular narrative politics of agency, subjectivity, and individualism. Such narrative politics privilege conceptual stability that pins down the molecularized actors on the eon-spanning evolutionary trajectory.

Several factors contribute to this prominence of the molecular evolutionary narrative as a textual strategy. First, as we have seen, the molecular evolutionary narrative produces the kind of coherence and completeness that evolutionary narratives focusing on species or organisms rarely achieve. Through the evocation of a historical scope and culturally familiar behaviors and practices, the minute interior world emerges as both (nearly) eternal and human-like, thus challenging the natural laws that govern our own mortal lives. Second, the hype surrounding the gene and DNA in popular culture makes the molecular evolutionary narrative an attractive tool both to those engaged in making science accessible to non-specialist audiences, and to those contemplating the ethical and philosophical issues arising from genomics. At the same time, the molecular evolutionary narrative reinforces the very hype surrounding genomics and visual technologies.

Third, the molecular evolutionary narrative's tendency to confirm our assumptions about the naturalness of cultural practices renders all the postmodern fuss about the volatility of gender and sexuality as either misguided or simply irrelevant. In this sense, the safe anthropomorphism that the molecular evolutionary narrative produces provides a sense of clarity and stability. Finally, the popularity of the molecular narrative may also arise from its valorization of science. If truth is located in the molecular and microscopic, science and scientific technology emerge as the only appropriate way of approaching it. This representation of science and technology caters to cultural anxieties about the vastness of all that we still do not understand by providing a sense of relief, security, and hope. This is the case even with Sykes' alarmist popular science, since he represents molecular biology as the only enterprise potentially able to lift Adam's curse.

As a textual device and an organizing structure, the molecular evolutionary narrative can produce extremely compelling stories of gender and sexual relations, the boundary between nature and nurture, and the role of science as a social and cultural authority. The next chapter shifts the focus of analysis to the narrative level of organisms and asks what happens when the sexual relationships between men and women are imagined within a framework informed by the idea of the molecular as the location of agency and humans as evolutionary puppets. Focusing on assumptions of innate infidelity in popular discourse, the chapter traces structural similarities between evolutionary and romantic narratives and explores the constitutive role of romance in the narrative logic of evolution.

Notes

1 For discussion and examples, see Keller (1995), Nelkin and Lindee ([1996] 2004), van Dijck (1998), Keller (2000), or Roof (2007).
2 See van Dijck (1998) for the history of the popular reception of genetics since the 1950s. Highlighting cultural ambivalence about advances in science, van Dijck contests the view that the historical development of popular attitudes toward genetic research and technologies followed a neat linear trajectory.
3 For discussion of metaphor in genetics, see, Fogle (1995), Keller (1995), van Dijck (1998), Hedgecoe (1999), van Dijck (2000), Ceccarelli (2004), Waugh (2005), and Roof (2007).
4 Schiebinger connects gendered anthropomorphism to eighteenth-century debates about gender differences and the proper role of women: "The new botanical sciences thus went hand in hand with the making and remaking of sexuality in the Enlightenment. Sexual images were prominent in botanists' language at the same time that botanical taxonomy recapitulated the most prominent and contested aspects of European sexual hierarchy" (Schiebinger [1993] 2004: 39).
5 The idea of genes as life-giving agents instead of mere parcels of inherited material can be traced to the 1944 publication of Erwin Shroedinger's *What Is Life*. See, for instance, Keller (1995) and Keller (2000). However, it was not until Dawkins that the idea of genes as the primary actors became the default position in evolutionary biology.
6 José van Dijck (1998: 94) makes a similar argument.
7 Men, of course, also carry only one X-chromosome, but since women have two X-chromosomes, the X-chromosome has benefited from the shuffling of DNA and has therefore presumably maintained its robustness.

4 The narrative attraction of adulterous desires

In 1972, Harvard sociobiologist Robert Trivers published an article entitled "Parental Investment and Sexual Selection." In this landmark paper, Trivers argued that differences in male and female sexual behavior can best be understood by examining how much energy (calories) and time each parent invests in an offspring. According to Trivers, "[i]ndividuals of the sex investing less will compete among themselves to breed with members of the sex investing more, since an individual of the former can increase its reproductive success by investing successively in the offspring of several members of the limiting sex" (Trivers 1972: 140). Building on A. J. Bateman's idea of eggs as energy consuming and sperm as cheap, Trivers argued that these differential parental investment strategies ultimately arose from "the initial disparity in size of sex cells" between males and females (Trivers 1972: 142). Translated into what Trivers took as the human heterosexual dynamic, this gametic disparity meant that "a copulation costing the male virtually nothing may trigger a nine-month investment by the female that is not trivial, followed, if she wishes, by a fifteen-year investment in the offspring that is considerable" (Trivers 1972: 145). Seen from the male's point of view, on the other hand, "[i]f there is any chance the female can raise the young, either alone or with the help of others, it would be to the male's advantage to copulate with her" (Trivers 1972: 145). As a result, "one would expect males of monogamous species to retain some psychological traits consistent with promiscuous habits" (Trivers 1972: 145). According to Trivers, then, the cultural double standard was carved in the most foundational of all cells, the gametes. At the same time, sexuality emerged as a battle between the sexes taking place in the interiority of the bodies and thus beyond the reach of culture.[1]

Trivers' theory of parental investment occupied a key place in the sociobiological enterprise that arose in the next few years. It was Trivers' (and Bateman's) hypothesis that made E. O. Wilson assert that not only do "[t]he consequences of this gametic dimorphism ramify throughout the biology and psychology of human sex" but "[t]he resulting conflict of interest between the sexes is a property of ... also the majority of animal species" (Wilson 1978: 124). With the publication of Dawkins' *The Selfish Gene* in 1976, this dictum became increasingly situated within an explicitly genetic framework, so that the binary differences between gendered gametes emerged as effects of genetic aspirations. Rewritten in Dawkinsian molecular vocabulary, the principle of gametic discrepancy appeared as a confirmation that natural selection favors "genes that say 'Body, if you are male leave your mate a little bit earlier than my rival allele would have you do, and look for another female'" (Dawkins [1976] 1999: 147). Crucially, this transfer of narrative activity across the behavioral, the cellular, and the molecular was underwritten by what some scholars have called the neo-Darwinian "reproductive imperative" (Terry 2000; Burns 2002). Positing competitive reproduction as the driving force of evolution, this adaptationist logic assumes that almost any behavior is an outcome of reproductive ambitions. At the same time, it also renders these same behaviors the very engine of further reproductive triumphs.

This view of humans as seeking to maximize their reproductive fitness through cheating and manipulation has become a prominent feature of today's popular discourses of gender and sexuality. In particular, evolutionary psychological accounts of human evolution that arose in the 1990s have repeatedly portrayed promiscuity and infidelity as the ultimate expression of the reproductive imperative. Such evolutionary stories about infidelity have proved extremely compelling as cultural narratives. For example, the portrayal of male desire toward young, slim, and big-breasted women as genetically programmed and thus presumably irresistible appears across the popular cultural spectrum from prime-time television series and Hollywood cinema to advertising and magazine articles. This portrayal resonates closely with popular assumptions about gender relations as a biologically determined heterosexual battlefield (Lancaster 2003; Lancaster 2006; Roof 2007; Schell 2007; McCaughey 2008).

In popular science texts, this evolutionary infidelity narrative is often spelt out in titles. Helen Fisher's *Anatomy of Love: A Natural History of Mating, Marriage, and Why We Stray* (1994), Robin Baker's *Sperm Wars: Infidelity, Sexual Conflict, and Other Bedroom Battles* ([1996] 2006), and Tim Birkhead's *Promiscuity: An Evolutionary History of Sperm Competition* (2000) are illustrative examples. Echoing Trivers' vision of gametic plotting, these texts typically proceed from a simple statement of a clash between male and female reproductive interests, synecdochically represented by the endless multitude of minuscule sperm and the large but rare eggs, to an extended speculation of how evolutionary mating strategies are visible in today's gendered sexual norms and practices. Reflected in chapter titles like "Polygamy and the Nature of Men" and "Monogamy and the Nature of Women" (Ridley [1993] 2003) or the wonderfully euphemistic "What Women Want" and "Men Want Something Else" (Buss [1994] 2003), this gametic gender dichotomy organizes the texts' discussions of specific instances of behavior, which are usually reduced back to the reproductive imperative.

One of the most striking features of this discourse of evolutionary infidelity is its insistence on the genetic hard-wiring of human emotions. As Roger Lancaster puts it: "Supposedly, the evolutionary effects of natural selection on the endocrine system have rendered us 'prewired' to have certain feelings, and (non sequitur) to express these emotional dispositions in certain pre-given institutional forms" (Lancaster 2003: 208). In particular, the evolutionary infidelity narrative has moved romantic love from the pedestal on which it has historically often been placed and challenged its cultural status as a unique, near-transcendental experience. In evolutionary psychological discourse, romantic love is viewed as a mere trick of our selfish genes, which exists only to improve our reproductive fitness. In other words, the feeling of love makes us copulate with the right kind of person in the right circumstances and stick with that person for the period of time that most likely leads to optimal reproductive success. As feminist psychologist Angie Burns observes, if "love is grounded in sexual and reproductive imperatives, it becomes explainable in biological and sociobiological or evolutionary terms as part of 'mate selection' and 'parental investment'" (Burns 2002: 150). This challenge to the "realness" of romantic love has been echoed in the media. The *New York Times*, for example, reported in April 2007 on a genetically coded "program for romantic attraction that makes people fixate on specific partners" (Wade 2007). In January 2008, *Time* magazine published a special issue on "The Science of Romance," subtitled "Why We Need Love to Survive," which included an article on why "Romance Is An Illusion" (Zimmer 2008).

This chapter explores how the highlighting of infidelity and the downplaying of romantic love in evolutionary discourse has altered the cultural understanding of evolution as a foundational narrative. The chapter focuses on the organizing logic of what could be called the *evolutionary infidelity narrative*, the narrative of the emergence, persistence, and primacy of promiscuous behavior within an evolutionary framework. In doing this, the chapter also examines the function of romantic love in the narrative dynamic of evolutionary transformation. In what follows, I first briefly discuss Darwin's view of promiscuity and his understanding of choice as the constitutive principle of sexual selection. I then turn to two popular science books, Matt Ridley's *The Red Queen* and Olivia Judson's *Dr. Tatiana's Sex Advice to All Creation*, through which I identify and compare two competing versions of the evolutionary infidelity narrative. The rest of the chapter provides a close reading of three novels, David Lodge's *Thinks …* , Alison Anderson's *Darwin's Wink*, and Jenny Davidson's *Heredity*, which all address the conflict between the ideas of prehistoric infidelity and all-conquering love. While diverse in their attitude toward evolutionary psychological claims of innate infidelity, the three novels all point to a parallelism between the structures of the infidelity and romantic narratives. This structural parallelism undermines the coherence of the evolutionary infidelity narrative by highlighting the possibility of reproductive failure. It also suggests that the narrative structure of romance may serve as a means of negotiating this narrative instability.

That choosy love

While the work of Trivers, Wilson, Dawkins, and other sociobiologists gave a scientific spin to the folk wisdom about men's adulterous inclinations, the idea of reproductive strategies as central to evolution echoed a much older work. In *The Descent of Man*, Darwin had posited mate choice as a significant evolutionary force through his theory of sexual selection. Darwin's insights into mating patterns were, however, initially dismissed by many of his contemporaries and followers, and it was not until the 1960s and 1970s that the theory of sexual selection was seriously reevaluated. It was precisely the controversial sociobiological project that reintroduced Darwin's work on sexual selection to wider audiences, making *The Descent*, as James Moore and Adrian Desmond put it, "the *locus classicus* of evolutionary psychology" (Moore and Desmond 2004: lvi). The continuities and contingencies between Darwin's and sociobiologists' ideas of sexuality are indicative of the changing meanings of movement and stability in evolutionary narratives of sexuality.

Whereas Trivers conjures up calculating parents who invest their resources in expectation of return, Darwin envisions in *The Descent* a natural kingdom inhabited by organisms that act like Victorian men and women. Notwithstanding the two texts' dramatically different discursive frameworks, both Darwin and Trivers understand individual choice as a crucial factor in the processes of evolution. For Darwin, sexual selection is the force that complicates and completes the work of natural selection by preserving traits that are not directly useful for survival. In fact, such traits may even seem to compromise an organism's prospects of survival, as in the case of ornaments that slow down movement while attracting predators. For Darwin, the mechanism of sexual selection consists of two types of "sexual struggle" (Darwin [1879] 2004: 684). The first type of struggle

> is between the individuals of the same sex, generally the males, in order to drive away or kill their rivals, the females remaining passive; whilst in the other, the struggle is likewise between the individuals of the same sex, in order to excite or charm those of the opposite sex, generally the females, which no longer remain passive, but select the more agreeable partners.
>
> (Darwin [1879] 2004: 684)

While there is a significant asymmetry in this scenario, as it is only males that compete among each other and "kill their rivals," both males and females are narrative agents with the power to choose. Whereas males fight for a chosen female, females actively select among several potential mates on the basis of their skill to "excite or charm." As many have noted, it was precisely this active role assigned to the female of the species that made *The Descent* so controversial among some of Darwin's contemporaries (Zuk 2002: 5–10; Hrdy 1999: 36).

I suggested in Chapter 1 that Darwin embedded his view of human sexuality within the discussion of the evolution of morality. In striking contrast to sociobiology, he equated true humanity with a rejection of bodily temptations and fleeting pleasures. In *The Descent*, Darwin consistently refers to promiscuity as a sign of a constrained evolutionary state, which he imagines as characteristic of primitive societies in accordance with nineteenth-century Western imperialism. For Darwin, one of the most alarming aspects of this presumed "promiscuous intercourse" and "licentiousness of many savages" (Darwin [1879] 2004: 655) is "the profligacy of the women" (Darwin [1879] 2004: 214–18), which he interprets as the ultimate signifier of a lack of evolutionary refinement. Western societies, by contrast, are portrayed in *The Descent* as having not only developed superior mental and physiological characteristics but also tamed the primitive inclination toward "[u]tter licentiousness" (Darwin [1879] 2004: 143). As a result, the presumably civilized Westerner

> cannot avoid looking both backwards and forwards, and comparing past impressions. Hence after some temporary desire or passion has mastered his social instincts, he reflects and compares the now weakened impression of such past impulses with the ever-present social instincts; and he then feels that sense of dissatisfaction which all unsatisfied instincts leave behind them, he therefore resolves to act differently for the future—and this is conscience.
>
> (Darwin [1879] 2004: 680)

The Descent constructs a dichotomy between fleeting, tempting urges and the stable, enduring social instincts, and portrays the latter as the basis of morality and ethics. While fully consistent with Victorian ideas of proper sexual behavior, this portrayal contradicts the reproductive premise of evolutionary change. For Darwin, it is the social instinct rather than the imperative to propagate hailed by sociobiologists and evolutionary psychologists that appears as fundamental to the human condition. At the same time, female sexuality emerges as the ultimate sign of evolutionary advance by positioning women at the extreme ends of the sexual spectrum: they are both the coy, sexually discriminate Victorian ladies and the horribly promiscuous native temptresses.

As many have observed, Darwin's portrayal of evolutionary development from ape to primitive to modern society in a landscape of moral elevation is somewhat confused. Roger Lancaster (2003: 88) notes that Darwin's idea of contemporary Westerners as having conquered their own primitive urges suggests paradoxically that Westerners had become more natural by rejecting their natures. As a result, "[m]odern Victorians' gender roles, family forms, and sexual practices are depicted as 'more natural' than those of either modern savages or past primitives—the very figures whose existence (at the supposed baseline of the human condition) lends rhetorical credence to the idea of 'natural origins' to begin with" (Lancaster 2003: 88). Furthermore, Lancaster argues, Darwin's account of the low evolutionary state of primitive societies undermines his thesis about the importance of female choice in sexual selection, as Darwin assumes that both Victorian women and females of other species choose their mates while in primitive societies it is the men who choose. As Lancaster points out, "[i]f primitive men long ago usurped the power of selection, then just how did Victorian women get it?" (Lancaster 2003: 87). Similarly, Darwin's depiction of the males of non-human species as adorned with ornaments to please females does not fit with his premise that in the Victorian society it is the women who are both "the beautiful, vain, preening creatures" and the ones who choose (Lancaster 2003: 87).

While these logical fallacies are indicative of Darwin's ideological commitments, there is a dimension in Darwin's theory of sexual selection that undermines his imperialist and sexist assumptions. Moore and Desmond argue that while modern sociobiologists and evolutionary psychologists emphasize "the natural selection of genes which have the effect of making females behave as if they were actively choosing their partners" (Moore and Desmond 2004: lvi), Darwin understands individual choice as a constitutive force in evolution. Furthermore, while Trivers, Wilson, and other sociobiologists consider reproductive sexuality as governed by preset rules and thus as a measurable and, to an extent, predictable evolutionary phenomenon, Darwin ascribes to desire a degree of unpredictability. As Elizabeth Grosz argues, Darwinian "[s]exual selection adds more aesthetic and immediately or directly individually motivating factors to the operations of natural selection; it deviates natural selection through the expression of the will, or desire, or pleasure of individuals" (Grosz 2004: 75). Furthermore, Grosz suggests, "there is something incalculable about the attraction to others, even in spite of attempted codifications of attractiveness in terms of evolutionary fitness" (Grosz 2004: 87). The logic of Darwinian evolution necessitates that what counts as attractive remains unfixed so that species may respond to the ever new varieties that evolutionary processes constantly produce. Sexual selection, then, engenders not only new anatomical and behavioral varieties but also new forms of beauty and allure.

As a productive force that pushes the evolutionary narrative in often unforeseen directions, sexual selection produces within the species a range of differences that are unnecessary for the sheer purpose of survival. In *The Descent*, Darwin portrays racial differences as precisely such accidental end-products of sexual selection, as some scholars have pointed out (Grosz 2004; Moore and Desmond 2004).[2] Darwin is explicit about this status of race as a shifting biological category. According to Darwin, "of all the causes which have led to the differences in external appearance between the races of man, and to a certain extent between man and the lower animals, sexual selection has been the most efficient" (Darwin [1879] 2004: 675). As sexual selection constantly gives rise to more varieties and, at the same time, to preferences for those varieties, "the differences between the tribes, at first very slight, would gradually and inevitably be more or less increased" (Darwin [1879] 2004: 665). Grosz maintains that this unending emergence of new modes of life destabilizes the very concept of race, rendering it simply "the long-term effect of increasingly intensified and diversified sexually selective criteria" (Grosz 2004: 85).

She interprets such criteria not in terms of survival but of visual appeal, as characteristics that "exert a *beauty*, an aesthetic force" (Grosz 2005: 24).

Grosz considers this primacy of sexual difference in Darwin's theory as an important contribution to feminist theory. While indeed more complicated and progressive than often assumed, Darwin's understanding of sexuality nevertheless seems to privilege a particular set of sexual practices. Most importantly, Darwin's portrayal of sexual selection as a productive force reinforces the reproductive logic that organizes his evolutionary narrative. Producing a dichotomy between gender and race, the primacy of sexual selection reduces racial differences to the evolving aesthetic economy of reproductive heterosexuality. As a result, racial differences emerge as products of the evolutionary narrative whereas gender differences are understood as its constitutive principle and narrative engine. This privileging of gender, however, does not necessarily trap the evolutionary narrative in a closed narrative economy. As we saw Grosz point out above, there is always "something incalculable" about desire and individual choice. While the primacy of gender over race renders all unaccountable variety within species as the product of sexual selection, it also places additional emphasis on the role of choice in the narrative dynamic of evolution. If there were no variety to choose from and no wish to choose differently from others, there would be no change, no evolution, no narrative.

This narrative logic places Darwin's portrayal of gender characteristics in a contradictory light. On the one hand, Darwin explicitly insists that evolutionary processes have generated "the present inequality between the sexes" (Darwin [1879] 2004: 631), since "it is the male which has been chiefly modified" (Darwin [1879] 2004: 625), while the human female has remained childlike, differing from men "in her greater tenderness and less selfishness" (Darwin [1879] 2004: 629). Consistent with Victorian gender ideologies, it is men who demonstrate "the greater size, strength, courage, pugnacity, and energy" as well as "greater intellectual vigour and power of invention" (Darwin [1879] 2004: 674), while it is women who have "sweeter voices" (Darwin [1879] 2004: 674) and show "the powers of intuition" and "imitation" (Darwin [1879] 2004: 629). On the other hand, Darwin argues that

> variations of the same general nature have often been taken advantage of and accumulated through sexual selection in relation to the propagation of the species, as well as through natural selection in relation to the general purposes of life. Hence secondary sexual characters, when equally transmitted to both sexes can be distinguished from ordinary specific characters only by the light of analogy.
>
> (Darwin [1879] 2004: 685)

Not only have characteristics been "accumulated" gradually through sexual selection but the processes of sexual and natural selection are often inseparable, so that a trait can be under the influence of both mechanisms and thus of use for both genders. Furthermore, what appears as fixed gender characteristics—the passive coyness of women and the energetic adventurousness of men—are always products of choice. Choice, by definition, relies on difference from past choices, since it is the possibility to choose differently that makes it a choice in the first place. By this logic, current gender characteristics and the choices they engender—no matter how fixed they appear—are also always constitutive of future genders and choices that will be different from the ones that are now. As the engine and the product of the evolutionary narrative, choice is both fundamental to the evolutionary narrative and yet too elusive to pin down to any fixed bodily, behavioral, or ideological position.

Promiscuous plotting

In stark contrast to Darwin's portrayal of sexual selection as interfering with the work of natural selection, the evolutionary narrative that emerged with sociobiology in the 1970s increasingly subordinated sexuality to a gene-centered, adaptationist logic of evolutionary change. As we saw in the previous chapter, Dawkins' gene's-eye view of evolution rejected the popular idea of sexuality as a mysterious force, representing it as calculative and strategic rather than wild and blind. Yet in continuation with Darwin's evolutionary narrative, the sociobiologists saw sexuality as the critical locus of evolutionary change that needed to be unraveled, explained, and accepted as a foundational truth about human nature. In the evolutionary psychological literature that arose in the early 1990s, this sociobiological portrayal of sexuality has produced narratives of modern men and women as driven by their plotting genes. This became evident, for example, in Bryan Sykes' account of the impressive (if eventually destructive) successes of the magnificent Genghis Khan Y-chromosome as the product of the emperor's unlimited infidelity. This emphasis on promiscuity and infidelity, however, is not simply a product of a shift in scientific discourses of sexuality. Assumptions of promiscuity also arise from the very logic of the evolutionary narrative. A closer look at the narrative of the evolution of adulterous desires in popularized evolutionary psychology will shed light on the function of infidelity in evolutionary discourse.

The evolutionary infidelity narrative we encounter in today's popular scientific discourse has two main storylines. The first (and normative) storyline builds directly on the narrative foundation laid by Trivers, Wilson, Dawkins, and other sociobiologists in the 1970s. Based on the thesis that males benefit from polygamy, this version of the evolutionary infidelity narrative portrays men as naturally promiscuous and women as naturally monogamous. Consequently, men are assumed to be cheating whenever they can while women are seen as searching for a reliable breadwinner endowed with good genes. This narrative is central to Matt Ridley's *The Red Queen: Sex and the Evolution of Human Nature* (1993), a book whose structure, style, and argument represent the mainstream of popularized evolutionary psychology. Confirming the common belief that men and women are fundamentally different, *The Red Queen* starts from the premise that "there are, in fact, two human natures: male and female" (Ridley [1993] 2003: 13), represented by "ardent, polygamist males and coy, faithful females" (Ridley [1993] 2003: 178).

The representation of gender differences in *The Red Queen* echoes closely Wilson's contradictory conjoiner of narrative transformation and permanence in *On Human Nature*. Like Wilson, Ridley understands evolutionary history as a constant interplay of adaptation and mutation. Yet in terms of gender, the gap between prehistory and modernity appears to be minimal, as Ridley's modern men, like the Pleistocene hunter, strive to "use wealth, power, and violence as means to sexual ends" (Ridley [1993] 2003: 206) and modern women, like prehistoric women, seek to "monopolize a man for life, gain his assistance in rearing the children, and perhaps even die with him" (Ridley [1993] 2003: 218). Ridley does occasionally complicate these gendered scenarios. He admits, for example, that a woman might want to have an occasional affair "with one well-chosen male" in possession of superior genes to boost her children's genetic makeup (Ridley [1993] 2003: 217). These exceptions to the binary logic of gender are, however, just that—exceptions. The way in which Ridley includes the mention of women's occasional polyandric preferences in the chapter entitled "Monogamy and the Nature of Women" suggests that these exceptions are ultimately subordinate to the principle of dichotomous sexuality.

The second version of the infidelity narrative arose in the 1980s after genetic tests revealed that in many species seemingly monogamous females give birth to offspring by multiple fathers.[3] Accordingly, this alternative infidelity narrative portrays both sexes as seeking opportunities for what is known in the technical literature as "extra-pair copulation." Olivia Judson's *Dr. Tatiana's Sex Advice to All Creation: The Definitive Guide to the Evolutionary Biology of Sex* (2002) is premised on this assumption. Written in the agony aunt format familiar from teen magazines, Judson's popular science book consists of "letters" from anthropomorphized animals—ranging from parasitic worms to mammals—and Dr Tatiana's answers to their sex-related worries. The text's attitude toward promiscuity and infidelity becomes evident in the case of a letter from "Crusading for Family Values in Louisiana":

Dear Dr. Tatiana,

My husband and I have been faithfully married for years, and we are shocked by what we read in your columns. As black vultures, we engage in none of the revolting practices you advocate so regularly, and we don't think anyone else should either. We suggest you champion fidelity or shut up.

(Judson [2002] 2003: 153)

In her reply to this amusingly moralistic letter, Dr Tatiana explains that, notwithstanding the sexual practices of black vultures, monogamy is "one of the most deviant behaviors in biology" (Judson [2002] 2003: 153). Elsewhere in the book, she consistently rejects female monogamy as "nonsense" (Judson [2002]

2003: 10) and declares that "in most species, girls are wanton" (Judson [2002] 2003: 9). This popular feminist assumption of equal promiscuity organizes the whole text, pitting fierce, competitive, and shamelessly promiscuous females against equally fierce, competitive, and shamelessly promiscuous males.

Despite the obvious differences in their representation of female sexuality, these two storylines are nevertheless variants of the same narrative. Driven by the same antagonistic rationale—outwitting the other sex is the key to reproductive success—both Ridley's and Judson's texts portray the two sexes as trapped in an oppositional logic. Invoking the Red Queen episode in Lewis Carroll's *Through the Looking-Glass,* Ridley envisions the males and females of every species as participants in an evolutionary chess game in which survival depends on the ability of competing genotypes to adapt to the ever-changing evolutionary environment and, most importantly, to the manipulative tricks developed by the other sex. This never-ending genetic sex war acquires capitalist and colonialist tones:

> For a man, women are vehicles that can carry his genes into the next generation. For a woman, men are sources of a vital substance (sperm) that can turn their eggs into embryos. For each gender the other is a sought-after resource to be exploited.
>
> (Ridley [1993] 2003: 175)

Judson's text embeds its narrative of infidelity in a similar discursive framework. The furious females and males that Dr Tatiana conjures are represented as caught in an "evolutionary arms race" (Judson [2002] 2003: 140). In this biological equivalent of the Cold War, the gains and losses of males and females are understood as mutually dependent, as "greater success for her often means less success for him" (Judson [2002] 2003: 7). Both texts, then, understand the evolution of infidelity as a narrative of ever-accelerating warfare between two irreconcilable sexual economies.

The reproductive imperative functions as the main narrative impetus in both versions of the narrative. That is, the evolutionary infidelity narrative is organized by the assumption that those traits that help organisms reproduce will survive to future generations and, conversely, that any surviving trait "must once have been (or must still be) the means to a reproductive end" since "[n]o other currency counts in natural selection" (Ridley [1993] 2003: 243). This adaptationist premise has two major implications for the evolutionary narrative. First, only events leading to reproduction can be structurally meaningful in the narrative dynamic of evolutionary change. My discussion of evolution as a foundational narrative in Chapter 1 evoked Robert J. O'Hara's argument that evolutionary history is always retrospective. According to O'Hara, evolutionary knowledge is the product of considerable historical distance, so that the "event" of an adaptation or speciation can be construed only when the event is already over and has been followed by further events (O'Hara 1988: 144–6; O'Hara 1992: 153). In the same way, the reproductive logic that organizes the evolutionary infidelity narrative is always a retrospective logic. If reproduction is indeed the only "currency" that "counts," as Ridley puts it, then acts that do not contribute to reproductive success cannot function as proper narrative events since they do not advance the progress of the evolutionary narrative—except perhaps as impediments that the master narrative needs to negotiate.[4] Second, the eon-spanning macro-level narrative of evolution— in this case, Ridley's endless chess game and Judson's perpetual arms race— comes to determine what we saw H. Porter Abbott call "a multitude of little stories of love and death" acted out by individual organisms (Abbott 2003: 147). Placed within this antagonistic master narrative, all individual deeds, desires, and decisions become portrayed as opportunistic maneuvers by dichotomously gendered organisms, who, driven by their presumably selfish genes, are always only trying to outwit each other. Searching to secure posterity, the reproductive narrative logic renders infidelity the preferred outcome of the little plots assigned to individual organisms.

The status of the reproductive narrative logic as a universal truth is assumed rather than explained in both texts. In *The Red Queen*, the differences between male and female sexual inclinations appear as frozen in time. While Ridley argues that the differences between the "two human natures" (Ridley [1993] 2003: 13) follow from the fact that *Homo sapiens* is a sexually (as opposed to asexually) reproducing species, this connection remains theoretical rather than temporal. In the same way, the connection the text claims to have located between human mating strategies and the evolutionary battle between host and parasite is analogous rather than strictly causal, and never quite explains the historical development from molecular-level disease resistance to the particular shapes human desire takes. Instead, the Red Queen chess game appears as an *a priori* rule. This representation echoes Lancaster's observation that evolutionary psychology tends to assume a "mythic time of evolution" (Lancaster 2006: 117), a past that is everywhere but yet not anywhere specific. As Lancaster notes, such freezing of the past tends to turn nature into a "great timeless mirror" that reflects the value systems of today's communities (Lancaster 2003: 96). Such stability also resonates with what Jay Clayton calls "genome time," the popular understanding of the genome as representing "perpetual present" (Clayton 2002: 33). Produced through a set of textual metaphors associated with the genome—the book of life, human library, DNA alphabet—genome time assumes unlimited possibilities of rewriting evolution, thus blurring the distinction between the past, present, and future (Clayton 2002: 33). This temporal ambiguity points to both resonances and tensions between the explanatory scopes of genomics and evolutionary theory, as many scholars have noted (Franklin 1995; Wald 2000; Turner 2007). It also undermines the linearity and one-way movement that is often associated with Darwinian evolution in general and evolutionary psychological narratives in particular.

Unlike Ridley, Judson imagines a point of origin for the ongoing adulterous arms race. Yet that point, too, is described in mythic terms that tend to freeze it into a timeless occurrence in the distant past. Consider Dr Tatiana's account of what she calls "the honeybee version of a chastity belt" (Judson [2002] 2003: 17), the phenomenon of the male honeybee leaving his genitals inside the queen after copulation in order to prevent the queen from mating with other males:

> Once upon a time, queens mated with only one male. Then a mutant queen appeared who mated with more than one. She was more successful at reproducing than her virtuous sisters, and the gene for multiple mating spread throughout the honeybee population. Then a male appeared who, by exploding, prevented the queen from mating again. Genes for exploding males spread throughout the population. In a counter-countermove, the queen evolved to block the male's advantage, either removing the plug herself or perhaps having it removed by the workers (this step would have happened swiftly, since any female who did not remove the plug would not have been able to lay eggs). Then males evolved their own counter-counter-countermove. And so on.
>
> (Judson [2002] 2003: 18–19)

In this comic-mythical portrait of originary monogamy ("once upon a time"), the female acts the part of Eve, initiating the fall that marks the beginning of the ever-accelerating, never-ending reproductive battle, epitomized in the endless ("and so on") chain of counter-counter-countermoves. The appearance of the first promiscuous female thus emerges as the foundational moment that alters the concept of time and the very mode of being, transforming peaceful but monotonous stability into a competitive dynamic of constant change. It is as if the emergence of female promiscuity revised ontology.

Finally, the reproductive logic that organizes the evolutionary infidelity narrative renders adulterous behavior reproductive not only in the sense that it produces more of the same by passing on genes to subsequent generations. Crucially, promiscuity is represented as being capable of engendering something entirely different that exceeds mere copying, such as new genetic combinations or adaptations. In Judson's text, "[c]onflicts of interest between males and females" produce "new weapon[s]" or "behavior[s]" that lead to the development of still more advanced weapons and still more effective behaviors, such as the skill to "manipulate and thwart" one's conspecifics (Judson [2002] 2003: 20). This productive logic is even more prominent in *The Red Queen*, as Ridley connects "the need to outwit and dupe" in the evolutionary infidelity game with the evolution of human intelligence (Ridley [1993] 2003: 192). He also maintains "that adultery may have played a big part in shaping human society" (Ridley [1993] 2003: 219), thus positioning adulterous desires as the engine that drives the narrative of human evolution toward its climactic endpoint, the emergence of modern culture.[5] This productive logic effectively counters the temporal ambivalence that the cultural discourses of genome time and the fantasies of rewriting evolution suggest.

The productive logic is reinforced by the fact that narrative itself is commonly understood as productive of new meaning. In her discussion of Darwin's narrative strategies, Gillian Beer observes that narrative as a structural pattern has a "tendency to align itself with a *purposive* explanation of the world it describes" (Beer [1983] 2000: 187; emphasis mine). In *The Poetics of DNA*, Judith Roof takes a somewhat stronger position:

> Narrative brings the assumption of certain values and cause-effect relationships with it. We assume, for example, that conflict will eventually produce something like a happy ending, knowledge, a victory, a product, a marriage, a child. This notion of production parallels our ideas of capitalist investment and payoff as well as our imagination of a heterosexual reproductive scenario.
>
> (Roof 2007: 18)

In the evolutionary infidelity narrative, infidelity is understood as initiating the evolution of something novel that reaches beyond what was before, thereby pushing the evolutionary narrative toward futurity. The productive logic associated with narrative thus buttresses the productive logic associated with infidelity. While reproduction itself provides the engine of the evolutionary narrative, it is infidelity that gives the evolutionary narrative a sense of ambition and invulnerability. Equated with the maximization of reproductive success, infidelity appears as the emblematic promise of narrative continuity.

Narrative masquerade

The cultural prominence of evolutionary psychological models of infidelity has engendered a number of fictional explorations of promiscuity and reproduction. The rest of this chapter examines three contemporary novels, David Lodge's *Thinks ...* , Alison Anderson's *Darwin's Wink*, and Jenny Davidson's *Heredity*, which all evoke the evolutionary infidelity narrative. This shift of focus allows me to examine the relationship between the evolutionary infidelity narrative and other cultural narratives, and thus address the charged interface between narrative structure and cultural context. As novels tend to be self-conscious about their own narrativity, they may point to structural ambiguities in the evolutionary infidelity narrative. At stake in such fractures is the sense of unending forward movement that distinguishes evolutionary narratives from many other cultural narratives, such as religious narratives.

Published in 2001, David Lodge's novel *Thinks ...* addresses evolutionary psychological theories of infidelity through its theme, plot, as well as narrative structure. Lodge's novel is framed as an exploration of the debates about the nature of consciousness in the 1990s, the decade envisioned as "The Decade of the Brain" by the first President Bush. Like many of Lodge's earlier campus novels, *Thinks ...* is built on a developing relationship between two characters that represent opposing viewpoints on the debated issue. This juxtaposition of contradictory views through two protagonists enables the author to educate his readers about intellectual controversies, as Robert P. Winston and Timothy Marshall (2002: 4) observe in the context of Lodge's 1988 novel *Nice Work*. In *Thinks* … Lodge's protagonists embody C. P. Snow's famous "Two Cultures" divide between the humanities and sciences. Director of the Centre for Cognitive Science at the imaginary University of Gloucester, Ralph Messenger is highly skeptical of both ends of the humanities: he constantly attacks traditional humanism (especially its belief in the "soul") while he simply "can't stand … postmodernists, or poststructuralists, or whatever they call themselves" (Lodge [2001] 2002: 228). A writer-in-residence at Ralph's university, Helen Reed represents the traditional humanist point of view, writing novels that, one of her reviewers comments, are "so old fashioned in form as to be almost experimental" (Lodge [2001] 2002: 340). In the course of the novel, the two characters engage in an extended debate about the development and meaning of human consciousness and emotions, which functions as the framework within which the novel's events are situated and given significance.

While Lodge's novel focuses on theories of consciousness, evolutionary infidelity discourse provides it with a strong narrative undercurrent. As in Lodge's earlier fiction—in the aforementioned *Nice Work*, for example—the central characters of *Thinks* … not only debate theories but also negotiate the possibility of an intimate relationship. And as in many of Lodge's novels, this potential relationship is adulterous, thus providing the narrative with a site for speculation on the evolutionary rationale for infidelity. Ralph's character functions as the main locus for these evolutionary psychological speculations. Throughout the book, Ralph, a married man with a number of previous extramarital affairs, tries to persuade the recently widowed and monogamous Helen to sleep with him. A chronic adulterer, Ralph reasons about his existence in a mock-Cartesian fashion: "*I think about sex, therefore I am*" (Lodge [2001] 2002: 293). He also repeatedly argues in terms of what things are "*for*, in evolutionary terms" (Lodge [2001] 2002: 69), thereby evoking the sociobiological foregrounding of adaptations and the idea of culture as an "unnatural" counterforce to our evolutionary nature. On the issue of adultery, he concludes: "[A]ddicted to sex, men are biologically programmed to want as much sex as they can get with as many women as they can get … only culture constrains our urge to copulate promiscuously" (Lodge [2001] 2002: 79). This portrayal of men as driven by their calculating genes follows closely the androcentric version of the evolutionary infidelity narrative we encountered in Ridley's *The Red Queen*.

This view of adultery as arising from evolutionary biology is reflected in other characters, most importantly Ralph's wife Carrie. Echoing her husband's views on sexuality, Carrie refers to female bodies as a set of "attribute[s]" that "got selected" (Lodge [2001] 2002: 207) and, Ralph reports to Helen, "knows most men are not one hundred per cent faithful to their wives" (Lodge [2001] 2002: 174). Carrie's portrait of male–female relationships is merely a variant of the never-ending battle of the sexes outlined in Ridley's and Judson's texts, as she states that whenever "the men have power and the women have youth and beauty, there's a trade-off. The men exploit their power to get sex, and the women exploit their looks to get promotion, or good grades, or just a good time" (Lodge [2001] 2002: 206). The view of infidelity as based on natural laws is given further support by the number of adulterous relationships the novel introduces. Apart from Ralph and Helen's evolving affair, the plot is built on a series of revelations about adulterous relationships, including, most importantly, that between Carrie and a family friend, Nicholas Beck, and the one that had taken place between Helen's late husband Martin and one of her current students at Gloucester.

What is interesting about Lodge's novel is that it articulates these evolutionary psychological arguments for a biological basis of adultery through the narrative structure of romance. In *A Natural History of the Romance Novel*, Pamela Regis identifies eight elements that she considers integral to any romance novel (Regis 2003: 30–38). According to Regis, romance novels typically open with the *definition of society*, understood as corrupt at the beginning but reformed in the end by the lovers' union. The plot also always includes such elements as the *meeting* between the heroine and hero, their *attraction* to each other, and the *recognition* of love (by both parties though not necessarily at the same point). The plot is complicated by the introduction of a *barrier* that prevents the lovers' union and a *point of ritual death* when the desired union seems absolutely impossible. In order to reach the conclusion required by the genre, romance novels include a *declaration* of love by both parties (again, not necessarily simultaneously) and the lovers' *betrothal*. While certain arrangements of these eight elements are more common than others, they can occur in any order (Regis 2003: 30–31), constituting, as Eric Murphy Selinger puts it, a kind of "neoclassical aesthetic" that "returns to first principles" (Selinger 2007: 312). While Regis focuses on the romance novel rather than romantic narrative in general, most of the elements she identifies are familiar from popular culture—especially from romantic comedies and television series, the latter of which recast them as an endlessly repeated pattern. These elements, then, reflect and reproduce popular ideas of what constitutes true love and what pattern (narrative structure) that love should follow.

Lodge's novel appropriates several of these elements. The actual courtship plot opens with the *meeting* of Helen and Ralph at a dinner party during which both characters feel instant (though still unacknowledged) *attraction* to each other. The two begin their professional and intellectual relationship—they discuss consciousness over lunch and Ralph shows the Centre to Helen—and in due course Ralph *declares* his (not love but) lust. The *barrier* in the relationship is Helen's feeling that an illicit affair with Ralph would offend her late husband's memory. The *point of ritual death* is reached when Helen finds out that her husband had been sexually involved with other women during their marriage. At this point, adultery appears as absolutely condemnable to her. The *barrier* is lifted when Helen discovers that Carrie, too, is involved in an extramarital affair. This is followed by Helen's *declaration* (again, of lust rather than love). While the novel's plot obviously lacks a *betrothal*—it is, after all, a story about adultery—these narrative elements associated with romantic fiction establish a narrative momentum that is immediately recognizable. The tension, evocation of impediments, and eventual climax and denouement that these elements produce resonate with our ideas of a romance plot. Yet by replacing a discourse of love with a discourse of lust and the infinite monogamous union with a finite affair, the infidelity narrative in Lodge's novel appropriates—one might even say masquerades as— romance rather than simply aligns itself with it.

Lodge's use of two characters that represent opposing viewpoints (on adultery as well as other issues) might suggest a turn toward an intellectual compromise at the end. This consolidation, however, never quite materializes. Even though the novel ends with Ralph's (false) cancer scare that, we learn, makes him "less assertive, more subdued, more middle-aged" and results in his "los[ing] his reputation for chasing women at conferences," this narrative conclusion does not question the evolutionary rationale for adultery outlined earlier in the text (Lodge [2001] 2002: 340). It merely suggests that the adaptation in question (infidelity) may have become maladaptive in modern life and needs to be subjected to self-imposed control—that is, it does not suggest that the adaptation would not exist. This interpretation is supported by the way in which the characters' intellectual exchanges are intertwined with the courtship plot. Initially against adultery and resistant to Ralph's persuasions, Helen gradually adopts elements from his evolutionary psychological discourse. Crucially, Helen's intellectual move toward Ralph's viewpoint coincides with her becoming Ralph's lover. For example, after exchanging a tentative kiss with Ralph in the backyard of his country cottage while his wife and children are waiting inside, Helen ponders her and Ralph's behavior in terms that connect adulterous deception with human evolution: "How adept at deception we human beings are, how easily it comes to us. Did we acquire that ability with self-consciousness?" (Lodge [2001] 2002: 105). In the same way, she uses words like "the sexual instinct" (Lodge [2001] 2002: 223) and, recalling their previous sexual encounter, imagines Ralph as "a Stone Age man taking his mate, short and sharp" out in the field (Lodge [2001] 2002: 261). The significance of Helen's sexual and intellectual yielding to Ralph is further underlined by the fact that this double move coincides with her awakening to the factuality of adultery as she discovers her late husband's affair.

While the intellectual differences between Ralph and Helen are never quite resolved—Helen retains the core of her humanist views after leaving Gloucester— the intertwining of courtship and intellectual debate at the level of narrative structure suggests that there is a profound if unarticulated agreement between the characters. In their discussion of *Nice Work*, Winston and Marshall observe that while the novel's lovers are separated at the end, "the intellectual coupling of Robyn Penrose and Vic Wilcox is every bit as 'real' as their sexual adventure in Frankfurt" (Winston and Marshall 2002: 10). In *Thinks …* this parallel appropriation of the courtship plot and the novel of ideas results in a similar intellectual and sexual coupling. As a result, the two levels at which the evolutionary infidelity narrative is invoked—the intellectual exchanges between the characters and the courtship acted out by them—appear as if reaffirming each other. It is as if by yielding sexually Helen also accepts Ralph's claims about the genetic basis of adultery. In this way, Lodge's novel subtly yet powerfully endorses evolutionary psychological discourse on adultery.

Despite the juxtaposition of evolutionary and romantic discourse in the popular media, then, there seem to be curious affinities between the two discourses. Although irreconcilable in terms of their views of human nature, the infidelity and romantic narratives turn out to rely on a similar narrative dynamic. Lodge's novel demonstrates that an appropriation of the underlying structure of romantic narrative may help deliver the evolutionary infidelity narrative in a culturally resonant form. Such structural familiarity may explain at least some of the popular appeal that the evolutionary infidelity narrative seems to have engendered. At the same time, the structural similarities between infidelity and romantic narratives also attest that infidelity is not a necessary product of the evolutionary narrative. Even if infidelity seems like the logical outcome of the reproductive premise that underlies evolution, the structural resemblance of the two narratives points to the possibility of telling the evolutionary narrative in terms other than those of adulterous genetic warfare. While assumptions of promiscuity and infidelity find support in the reproductive logic of the evolutionary narrative, such logic may also be able to accommodate other cultural vocabularies and narrative commitments.

Enduring the attraction

Alison Anderson's *Darwin's Wink: A Novel of Nature and Love* (2004) provides a contrast to Lodge's novel through its poetic, often melancholic tone and the way in which it juxtaposes genetically controlled infidelity and romantic love. At the level of plot, *Darwin's Wink* depicts the experiences of its two protagonists, American behavioral ecologist Fran and her Swiss assistant Christian, on an island off the coast of Mauritius, where they try to "abet the survival of the weakest" by saving the mourner bird from extinction (Anderson 2004: 163). Both characters are haunted by a sense of disappointment and failure: Christian is only just recovering from his experience as an International Red Cross delegate in Bosnia during the atrocities in the early 1990s and Fran has chosen the solitude of Egret Island to escape the false sense of community that her life in Berkeley provided. Both Fran and Christian have recently faced loss: Christian's colleague and lover Nermina disappeared in Bosnia during the war and Fran's Mauritian assistant and lover Satish drowned in a malevolent attack on a boat off the island. While the novel develops a mystery subplot that involves violent assaults on the birds, a ransacking of Fran and Christian's abode, and an attack on Christian on the dinghy, the novel's thematic focus is on the questions of survival, death, free will, and determinism.

The novel's plot, language, and characterization are all embedded in an evolutionary framework, as suggested, for example, by Fran's portrayal of California as a place where "34 million people competed ... for space and survival" (Anderson 2004: 83). Viewed from Fran's persistently Darwinian viewpoint in particular, all human activity appears as a mere variant of animal behavior.

The novel's treatment of genetic determinism and free will provides the textual site where the evolutionary infidelity narrative and romantic narrative become juxtaposed. Like Lodge's *Thinks ...* , *Darwin's Wink* represents infidelity as an expression of a genetically coded evolutionary adaptation. Most of the evolutionary speculations take place through the notes, thoughts, and comments of Fran, who acknowledges the primacy of the evolutionary reproductive imperative, even though she personally "despises promiscuity," a statement that echoes her own experiences of her ex-husband's adultery (Anderson 2004: 82). Consistent with the male-centered version of the evolutionary infidelity narrative advocated by Ridley, Fran believes that all men act on irresistible instinctual desires. When finding out that Christian is seeing a local young woman called Asmita, Fran writes in her journal:

> *Am I surprised? No. Male behavior pattern. Disappointed? Yes. I thought he was different. Even if I know the way a male—of any species—is supposed to behave, I always hope the male of the human species will use those specific faculties* which make him human *to change the more animal sides of his nature.*
>
> (Anderson 2004: 74)

This view of male sexuality as natural, ancient, and indisputable is not just Fran's particular evolutionary interpretation. Christian's experiences are described through similar vocabulary at several points in the text where Fran is not the focalizer. For example, when Christian discovers that Asmita's family does not accept premarital sex, we learn that "being a man he wants much more and fights his own self-control" (Anderson 2004: 50). We also find him wondering "[h]ow much of his desire belongs to him ... and how much to that purely genetic instinct implanted to assure the survival of the species?" (Anderson 2004: 50).

These assumptions about the naturalness of promiscuity and the irresistibility of the male sex drive are embedded in the evolutionary psychological discourse of reproductive fitness. Unable to conceive and left by her ex-husband "for a younger, fertile woman" (Anderson 2004: 197), Fran sees herself as "outside of the larger scheme. Of Nature's scheme, even" (Anderson 2004: 210). She also imagines herself as "missing a certain gene" because she does not wish to participate in the promiscuous mating games of human societies, and reasons about her own biological existence: "I am misadapted. In any other species I would not survive" (Anderson 2004: 110). Likewise, when she and Christian embark on a brief intimate relationship toward the end of the book, she describes their lovemaking in her journal as having no future because "*[h]e plants his seed in me, useless, fruitlessly*" (Anderson 2004: 209). This question of reproductive fitness finds further resonance in Fran and Christian's struggle to get the mourner birds to mate and reproduce. It is also echoed in the image of the child that appears as the seal of love stories throughout the novel: Fran's ex-husband now has two sons whose names sound like "a Romanov dynasty" (Anderson 2004: 19), Fran rejects Satish's pleas that they should have children right before his death, Asmita conjures up "a small nut-brown boy" while daydreaming (Anderson 2004: 101), and Christian imagines his and Nermina's unborn child as "holding a small bear, or a doll, or a piece of bread" (Anderson 2004: 254).

The evolutionary infidelity narrative is further reinforced through the text's repeated highlighting of the failure of communication and the prominence of lies in human relationships. Sometimes the connection between language and adultery is explicit, as when Fran reasons that "[p]erhaps lying is an evolutionary survival tool" since "adultery entails lying" (Anderson 2004: 218). This association between infidelity and lies finds further support in the novel's more general problematization of language. Again, it is Fran the Darwinian who imagines societal discourse as "words rising bubble-empty on the buggy night" (Anderson 2004: 67) and ponders "why man had invented language, when song was often more true" (Anderson 2004: 19). Evoked consistently throughout the text, this ineptitude of language is anticipated by one of the novel's two epigraphs: "Which language never lies? The language of animals." Represented as a traditional Mauritian riddle, the epigraph situates the question of lying—and, by implication, adultery—within the larger context of animal behavior. It is, then, human communication rather than communication in general that is remarkably inefficient and unreliable.

This communicative failure is given a further twist through repeated references to the idea of men and women as speaking different languages, an assumption made popular by John Gray's *Men Are from Mars, Women Are from Venus*, as suggested in Chapter 3. The communicative barrier between the genders is introduced early in the novel. When Christian arrives at Egret Island, the linguistic distance between him and Fran is contrasted with the immediate recognition of "a sympathy, perhaps a male thing" between Christian and Sean, the Irish leader of the preservation program on Mauritius (Anderson 2004: 64). Celebrating the hatching of two mourner-bird chicks with Christian and Sean, Fran feels that she's an outsider, only "pretending to be part of this conversation" (Anderson 2004: 66). Similarly, the brief encounters between Christian and Asmita are described in terms of a linguistic break that cannot be reduced to cultural difference (Asmita is credited as having studied in Switzerland). When Christian and Asmita kiss in Christian's car, "her words fade, become unintelligible, translated into his desire to place his lips against her skin" (Anderson 2004: 142), and when Christian leaves her later in the novel, Asmita's words are equated with "querulous calls of a stranger, speaking a language he does not know" (Anderson 2004: 152). This evocation of communicative failure has a twofold effect. First, it facilitates a metonymic transfer from gendered language to gendered nature, thus buttressing the idea of men and women as embodying different modes of existence and, consequently, as driven by different needs. As a result, men's presumed adulterous proclivities can function as a sign that separates what Ridley imagines as the "two human natures." Second, lies in heterosexual intimate relationships appear as inevitable, since the coexistence of two incompatible languages is likely to produce interpretative confusion and unfulfilled expectations.

Despite these repeated references to infidelity as a genetically programmed male trait, *Darwin's Wink* ultimately rejects any clear position on promiscuity. This ambiguity arises from the novel's narrative organization, which contrasts the speculations about the evolutionary function of polygamy with the two protagonists' experiences of what they consider as genuine love: Christian's relationship with Nermina, Fran's relationship with Satish, and, to an extent, Fran and Christian's brief sexual relationship. Compared to the complexities of these love stories, the other narratives of genetically driven encounters seem formulaic. Fran is well aware of this, as suggested by her description of a scene at a cocktail party she is attending. Witnessing the exchange of suggestive looks and bodily signs between a man and a woman, she compares the way the man is "leaning toward" the woman, "moving his head from side to side" (Anderson 2004: 109) and the woman is "flashing" her eyes and producing a rhythmic "throaty laughter" (Anderson 2004: 110) to "the courtship rituals of seabirds—albatrosses in particular—who bow and nod and engage in a hilarious dance of outstretched necks" (Anderson 2004: 109). Human mating rituals, the analogy suggests, follow a predictable pattern, leaving little space for originality.

Christian, too, appears to be aware of this predictability. In the course of the text, he grows increasingly suspicious of the romance plot Asmita is offering him, for her story just seems too straightforward. According to the plot Asmita envisions, their initial meeting is followed by a series of increasingly passionate encounters in his car, an engagement, a wedding, a child, and regular holidays in Switzerland. Even the narrative crisis that her plot establishes is a false one: she has lied to her parents that Christian has made love to her so that Christian would agree to marriage in order to save her honor. Crucially, Asmita's romance plot falls apart because the narrative crisis does not carry it. When Christian finds out about her deception, he rejects her romance plot as false, with the consequence that their narrative fails to build toward the final climax and instead collapses right at the crucial moment. Unlike Lodge's *Thinks …* , then, *Darwin's Wink* imagines the male urge to reproduce as unable to produce an adequate, structurally satisfying romantic narrative. What complicates this further is the fact that Christian and Nermina's relationship in Bosnia was—in a technical sense at least—adulterous, even though Nermina's husband had been missing for two years at the point when their story begins. In a curious reversal of the narrative politics of Lodge's novel, it is not the infidelity narrative that passes as a romantic narrative but the romance acted out by Christian and Nermina that masquerades as an infidelity narrative.

Compared to the reproductively motivated and genetically driven false plots, the novel's romances are full of surprise, tension, and narrative movement that exceed a mere straightforward procession of narrative events. To apply Regis' terminology, Christian and Nermina's story relies on an extended point of ritual death, as the events on the island take place when Nermina has disappeared and their union seems impossible. This ritual death eventually produces an emotionally satisfying ending, as the novel closes with Christian's discovery that Nermina is alive. The narrative dynamic of Fran and Satish's story, on the other hand, involves a strong element of surprise that accompanies Regis' element of *recognition*, present, for example, when Fran, having caught herself thinking about love and Satish, tries to tell herself to "Accept it, love's an illusion" (Anderson 2004: 75). Sudden recognition is also central to Fran and Christian's later intimate relationship, even if their relationship is one of sensual and emotional discovery rather than love. Their first kiss is described as happening "slowly, restrained by astonishment, This cannot be happening, why suddenly now, why us?" (Anderson 2004: 203–4), and they constantly feel "that they have been caught out by the unexpected" (Anderson 2004: 214). Unlike mere genetically driven coupling, the novel suggests, love is unpredictable. What distinguishes the romantic narrative from the infidelity narrative, then, is its reliance on the unforeseen as its narrative engine.

Finally, the novel ascribes to love the power to resolve the communicative failure between men and women. Contrary to the reproductively motivated, incompatible syntaxes of male and female sexualities that evolutionary psychology often assumes, love in *Darwin's Wink* can abridge the grand canyon of sexual difference that Bryan Sykes imagines in *Adam's Curse*. For example, Christian contrasts Asmita's insistent questions about whether he loves her with the wordless understanding between him and Nermina: "Did she ever ask him that question, Nermina? Did it need asking, when every gesture, every day they were alive was like a confirmation of their love for each other?" (Anderson 2004: 105). The communicative break between Fran and Christian is also mended through what is represented as an intuitive connection, as Fran is caring for the feverish Christian injured in the attack on the dinghy:

> Later she will not remember how she learned the story, whether he told it to her in his gravelly voice, or she went with him to the source of his dreams. When she asks him, he will say, I never told you, you're imagining things.
>
> (Anderson 2004: 176)

It is through sensation and intuition rather than the treacherous human language that men and women can communicate. This is further supported by the novel's emphasis on the reliability of animal communication, especially the harmonious song of the mourner birds, which, crucially, are also known for their "obvious fidelity" that lasts a bird's lifetime (Anderson 2004: 157).

Despite the persuasive pull of romantic narrative, *Darwin's Wink* nevertheless remains skeptical about the possibility of love to survive in what it portrays as a genetically driven world. The final reunion of Christian and Nermina, for example, is a matter of pure chance in the chaotic circumstances of a war zone. This highlighting of chance resonates with the representation of evolution as guided by what Fran refers to as "the stochastic factor, the random determinant" (Anderson 2004: 13). This unpredictability implies that love is in danger of being drowned amongst the abundance of promiscuous and reproductively calculative couplings. The emphasis on chance also echoes Darwin's understanding of individual choice as intrinsic to sexual selection and constitutive of evolution. Consistent with Darwin's view of choice, love in *Darwin's Wink* is able to push the narrative—if not quite evolution itself—toward unions that defy differences of age, ethnicity, and cultural heritage and, in the case of Fran, simply reject the role of reproduction as a narrative engine. This dissimilarity between the evolutionary infidelity narrative and the romantic narrative suggests that the latter cannot be reduced to the logic of the former. The contrast between the formulaic pattern of genetically motivated mating and the unexpected twists of romance implies that there is always something that falls outside the logic of the evolutionary infidelity narrative. Even when there is a formal similarity between the events of the two narratives—two individuals court, mate, and reproduce—the evolutionary infidelity narrative lacks the narrative tension produced by desire, emotion, sensation, and longing. I return to this question of narrative "surplus" in Chapter 5.

Structural resistance

Jenny Davidson's 2003 novel *Heredity* shares key thematic concerns with Lodge's *Thinks ...* and Anderson's *Darwin's Wink*. Like the other two novels, *Heredity* refers to developments within the sciences in the 1990s, although Davidson's main focus is on reproductive technologies. As in the other two texts, evolutionary psychology appears as a significant part of the discursive framework within which the novel's sexual relationships are portrayed. In terms of plot, *Heredity* tells the story of a young American woman, Elizabeth Mann, who becomes the London correspondent for an American budget travel guide. In London, Elizabeth meets Gideon Streetcar, a gynecologist and fertility specialist who idolizes her father, the famous Dr Mann of Yale, his earlier mentor. As a protagonist, Elizabeth stands clearly apart from both Lodge's conventional Helen and Anderson's cynical Fran. Spontaneous and uninhibited about sex, Elizabeth has initiated an adulterous affair with the married Gideon by page 15, and the rest of the novel portrays explicitly and nonsentimentally a series of sexual encounters between the two characters.

The narrative of Gideon and Elizabeth's illicit affair is paralleled by a historical narrative set in eighteenth-century London. Early in the novel, Elizabeth accidentally discovers the journals of Mary Wild, wife of Jonathan Wild, the notorious eighteenth-century criminal and self-acclaimed "Thief-Taker General of Great Britain and Ireland" (Davidson 2003: 39). In the journals, Mary Wild records—often in sexually explicit detail—her life first as Jonathan Wild's housekeeper, then as his wife, and finally as his son's lover. Rather than a simple narrative of an adulterous relationship, *Heredity* is an account of Elizabeth's obsessive search for the truth about Jonathan Wild, who—unlike Gideon with his asthma inhaler and low sperm count—stands for the mythic figure of the brutish but sexually irresistible evolutionary alpha male. This obsession culminates in Elizabeth's determination to give birth to Jonathan Wild's clone. Elizabeth persuades Gideon to assist her, but desperate for a child and knowing that the procedure is technically unfeasible, Gideon reimplants in Elizabeth's uterus one embryo that carries Wild's DNA and two embryos produced with his own sperm. The novel ends with Elizabeth's realization that she is pregnant with Gideon's child instead of Wild's clone.

Evolutionary psychological discourse enters the text through Gideon and Elizabeth's dialogue and Elizabeth's first-person account of the novel's events. Like Lodge's Ralph—and, to an extent, Anderson's Fran—Gideon delivers evolutionary psychological dictums about men and women's innate desires. Evoking the constitutive role of reproduction in the evolutionary narrative, Gideon states, for example, that every woman has a "right to pass on her genetic material to her offspring, to perpetuate the family line in her descendants" (Davidson 2003: 27). He also echoes the evolutionary understanding of relatedness as a matter of genetic inheritance and refuses to consider adoption as a way to build a family: "You think I'd be willing to raise a child who doesn't carry my genes? I can't afford it. Generations of dead Jews, all depending on me to propagate. I'd do anything for a child of my own. Anything" (Davidson 2003: 27). He projects this "pull of heredity" (Davidson 2003: 27) not only on his inherited membership in an ethnic and religious minority (the Jewish expatriates in the UK) but also on the evolutionary reproductive imperative itself, asserting that "[e]volutionary biology has shown pretty conclusively why it is that stepfathers kill their children at a higher rate than natural fathers" (Davidson 2003: 218). For Gideon, reproduction is, to echo Ridley again, the only currency that counts. Moreover, while Fran in *Darwin's Wink* views this reproductive imperative within the larger scheme of evolution, Gideon sees it only in terms of individual rights.

At several points in the text, Elizabeth adopts similar popular evolutionary discourse. After their first sexual encounter in Gideon's office, for example, she likens her and Gideon to "wolves on the Discovery Channel" (Davidson 2003: 14), and elsewhere she fluently employs technical terms like "extra-pair paternity" (Davidson 2003: 126). Throughout the novel, Elizabeth acts out the feminist version of the evolutionary infidelity narrative we encountered in Judson's popular science. She not only initiates the adulterous affair with Gideon but, like Dr Tatiana's happily promiscuous females, also has casual sex with other men when occasion permits. The historical subplot reflects this representation of female sexual initiative: like Elizabeth, Mary Wild actively proceeds to seduce the then married Jonathan Wild, and later initiates the adulterous relationship with Wild's son. Moreover, both characters seem to be driven by uncontrollable sexual urges: Elizabeth is often "in a state of acute sexual arousal" (Davidson 2003: 214) and decides to "take what [she] can get" (Davidson 2003: 215). Likewise, upon first meeting Jonathan Wild, Mary Wild "could hardly take [her] eyes off the man, and ... thought [she would] need little persuading to consent even to [her] own ruin" (Davidson 2003: 46). Furthermore, Elizabeth and Gideon's affair is always more about lust than love. Right before seducing Gideon, for example, Elizabeth notes that she does not "like him much" (Davidson 2003: 10), and later she states that relationship is "a word I hate" (Davidson 2003: 108). Even her reluctant admission that she might love Gideon ("I guess so ... Yes.") later in the novel is never really substantiated by anything she does, says to Gideon, or reports to the reader (Davidson 2003: 189).

Despite these references to evolutionary psychological discourse, *Heredity* does not advocate an adaptationist view of infidelity. First, Elizabeth's attitude toward evolutionary explanations of gender and sexuality is ambiguous at best, as suggested by her listing of Gray's *Men Are from Mars, Women Are from Venus* as one of "the more offensive classics of the self-help movement" (Davidson 2003: 112). She also dismisses Robert Wright's *The Moral Animal*— a well-known work of evolutionary psychology—as "reactionary polemic" (Davidson 2003: 208). Second, while the novel marginalizes romantic discourse, it also rejects the narrative dynamic of romance so central to Lodge's novel. Significantly, Davidson's novel also differs from Anderson's juxtaposition of the structural monotony of the evolutionary infidelity narrative and the dynamic appeal of the romantic narrative. Like *Darwin's Wink*, *Heredity* represents the evolutionary infidelity narrative as lacking the productive element of surprise and thus what could be called structural stamina. Davidson's novel, however, takes this critique still further by denying the infidelity narrative any resemblance with romantic narrative. Furthermore, instead of substituting romance for the infidelity narrative as *Darwin's Wink* does, *Heredity* rejects the very logic of romantic narrative.

Unlike Lodge's and Anderson's novels, *Heredity* lacks all of the key elements that Regis identifies in the popular romance novel. There is no narrative trajectory in Elizabeth and Gideon's relationship: their sexual encounters appear as a linear sequence that lacks development and thus tension, climax, or denouement. There is no structurally significant impediment in the relationship either. Since the affair is not built primarily on love but lust, Gideon's marriage does not represent what Regis calls a *barrier*. Similarly, Elizabeth and Gideon's *attraction* to each other is reduced in significance by the fact that both have previously been engaged in other similar relationships, and there is no climactic scene of *recognition* or *declaration*. Moreover, the element of *recognition* has been displaced into the female *Bildungs* narrative—Elizabeth's search for direction in life—that emerges from the structural and thematic intertwining of the modern and historical plots through Elizabeth's obsessive interest in Jonathan Wild. This *Bildungs* narrative, however, departs significantly from the conventional *Bildungs* narrative, since Davidson does not use romantic love as a life lesson that contributes to the main character's development as many *Bildungs* narratives do. In fact, we do not even recognize the novel as a *Bildungs* narrative until the very end. In a further act of displacement, this educative function is projected onto the cloning plot. It is Gideon's professional betrayal in their joint cloning project and Elizabeth's own false beliefs about genetics that lead to her subsequent pregnancy, forcing her to face her situation and, we assume, reject the fantasy world she has invented.

Read in the context of Lodge's narrative strategies, Davidson's refusal to rely on the narrative dynamic of romance is also a refusal to let the evolutionary infidelity narrative fully develop and gain textual control through the structures of romantic narrative. Such narrative resistance also subtly questions the self-evidence of the reproductive imperative as the logic driving evolutionary change. If the evolutionary infidelity narrative is underwritten by the imperative to enhance reproductive fitness, it is a curiously impotent narrative, since it seems unable to engender either a plentitude of offspring or proper narrative development. This structural resistance is further supported by the novel's deployment of infertility imagery. Both Gideon and Elizabeth's father are infertility experts, and many of the characters are unlikely to reproduce without technological assistance: Gideon's wife Miranda is unable to conceive, Gideon has a low sperm count, his mother Clara had difficulty getting pregnant, and Jonathan Wild's first wife dies in childbirth when finally pregnant. Furthermore, Elizabeth's ambiguous relationship with her father and her determination to have "a child that would have nothing of myself in it" suggests that the title *Heredity* refers as much to cultural and social forms of relatedness as to genetic inheritance (Davidson 2003: 269).

Through its rejection of romantic narrative and its evocation of infertility imagery, Davidson's novel parodies the exclusive focus on reproduction in the evolutionary infidelity narrative. By doing this, the text also suggests that the reproductive logic underlying the evolutionary infidelity narrative relies on a rather awkward precondition: the evolutionary narrative can proceed only if reproduction succeeds. By thematizing infertility, the novel implies that the evolutionary infidelity narrative depends on a constant denial of the possibility of reproductive failure. Interestingly, Davidson's novel is not alone in articulating such cultural anxieties about futurity. As we have seen, both Simon Mawer's *Mendel's Dwarf* and Alison Anderson's *Darwin's Wink* invoke the understanding of infertility as a fatal failure in the Darwinian paradigm. If the evolutionary narrative is based on reproduction, these texts suggest, infertility signifies the end of all narrative. *Heredity* differs from these other texts in the extent to which it takes this critique. Through its exceptionally thorough structural resistance to narrative in general and romantic narrative in particular, *Heredity* explicates the unresolved tension between reproductive success and reproductive failure that underlies the evolutionary infidelity narrative.

Infertile plotting

Lodge's, Anderson's, and Davidson's novels suggest that fictional texts may contest the reproductive evolutionary logic by reworking the relationship between the infidelity and romantic narratives. But what exactly is the narrative dynamic through which such a challenge may emerge? What is the relationship between discourses of infidelity, infertility, and the narrative structure of romance? In order to address these questions, we need to take a closer look at the reproductive narrative logic itself.

In "Homosexuality and Narrative," literary scholar Dennis W. Allen (1995) analyzes the inherent precariousness of what he calls the narrative of heterosexuality. Using Homi Bhabha's notion of the nation as produced and maintained through two narrative modes, the pedagogical and the performative, Allen suggests that our understanding of the primacy of heterosexuality is produced through a similar "double narrative movement" (Allen 1995: 615). Building on Judith Butler's work on the performativity of gender, Allen argues that the performative narrative of heterosexuality relies on the constant repetition of everyday acts and utterances. The pedagogical narrative of heterosexuality, on the other hand, has two alternative storylines that both insist on a connection between heterosexuality and reproduction, and, by extension, between heterosexuality and the origins of human existence:

> Understood in religious terms, heterosexuality is seen as the literal originary moments of creation in the Garden of Eden and of the fall that inaugurated sexual reproduction (the resulting status of homosexuality then being summarized by the catchphrase, "God created Adam and Eve, not Adam and Steve"). Alternatively, when understood "scientifically," heterosexuality is seen as indistinguishable from a transcendentally conceived sexual biology, and its originary status is displaced from a particular "historical" moment into an image of temporality itself: the infinite regress of the generations.
>
> (Allen 1995: 616)

Allen's second pedagogical narrative corresponds with the Darwinian evolutionary narrative. This narrative hides the precariousness of its central claims—the falsely assumed connection between all heterosexual activities and reproduction, for example—behind the powerful image of "the infinite regress of the generations," thereby naturalizing a particular set of sexual practices. By doing this, it reinforces what queer theorist Lee Edelman has called "reproductive futurism," a logic that posits reproduction as the single gateway to posterity (Edelman 2004: 2). In this sense, evolution, even when concerned with the past, is always a discourse about the possibility of a future.

According to Allen, what makes the pedagogical narrative of heterosexuality unstable is that "unlike national narratives, whose originary moment(s) are historical, its emphasis on reproduction, at least in the scientific version of this narrative, means that it must be exemplified, endlessly repeated, by individual performance" (Allen 1995: 618–19). Consequently, failures of performance may engender "conceptual instability" (Allen 1995: 619). The retrospective nature of evolutionary knowledge tends to mask this instability, as the fact of evolutionary descent is itself evidence of reproductive success. In this sense, evolution is, as Robert O'Hara points out, history written from the point of view of the winners, of those who have achieved the authoritative position from which to look back (in the case of evolution, *Homo sapiens* as opposed to any other species) (O'Hara 1992: 143). However, this retrospective aspect of evolutionary knowledge also implies that the continuation of the evolutionary narrative in the future may be fantasized but never guaranteed. Thus when Ridley writes that women are men's "means to genetic eternity" in *The Red Queen*, the never-ending future he evokes is essentially an illusion (Ridley [1993] 2003: 244). Similarly, Judson's "and so on" arms race between the honeybee queens and drones can take place only if reproduction succeeds. As the next chapter will argue, it is in such visions of narrative futurity that evolutionary narratives become perhaps most fundamentally entangled with ideology.

Furthermore, the precariousness of the reproductive narrative logic that Allen identifies shares affinities with Darwin's understanding of the unexpected as a characteristic of sexual selection. As we saw earlier in this chapter, Darwin does not question the primacy of reproduction in evolution. He does, however, propose that what gets reproduced may not be what is expected. Premised on the presence of choice, sexual selection may privilege new varieties while threatening the continuity of what seemed like a stable set of traits in the population. Even though Darwin represents male and female sexual behaviors as largely fixed, the emphasis he places on transformation and unpredictability logically defies this stability. When applied to the question of infidelity, this logic of continuous change suggests that promiscuity and infidelity may indeed be favored by evolution at a particular historical moment and in a particular historical environment. Such strategies, however, are never stable, and promiscuity may simply cease to be the best means of enhancing an organism's reproductive fitness. There is, then, a fundamental contradiction in the evolutionary psychological representation of evolution as movement toward something new and infidelity as written in the genome. This is, of course, the same contradiction that undermines Wilson's portrayal of evolution as a constantly developing foundational narrative and the genders as prisoners of a prehistoric biology in *On Human Nature*.

I suggested above that the infertility imagery in Davidson's *Heredity* can be read as symptomatic of the inherent precariousness of the evolutionary narrative in general and the evolutionary infidelity narrative in particular. As we saw, infertility imagery is also central to Anderson's *Darwin's Wink*, whose main narrative is framed as a struggle to save a species by trying to prevent reproductive failure. Infertility imagery is, however, practically absent in Lodge's

Thinks ... . In fact, Lodge's novel does not seem to be at all concerned with reproduction, as all its main characters are in their forties or fifties and there is no mention of the possibility of future progeny.[6] Yet this absence of reproductive concerns does not seem to undermine the reproductive underpinnings of Lodge's evolutionary infidelity narrative. I argued above that the way in which *Thinks* ... appropriates the narrative dynamic of romance produces a curious sense of narrative confidence. It is as if the imitation of romantic narrative made the text invulnerable to structural challenge. Consequently, it is pivotal to understand the productive dynamic of romantic narrative through which this effect is produced.

While Lodge's appropriation of the romance plot operates as a narrative disguise for the evolutionary infidelity narrative, it also serves as a means of insisting on narrative continuity against the risk of discontinuity inherent in the reproductive imperative. This is because romance as a narrative dynamic produces a sense of narrative continuity, as feminist literary scholarship has demonstrated. Exploring a wide range of texts from popular romance novels and romantic comedies to female *Bildungs* narratives and fictional narrative in general, feminist scholars have long argued that the narrative dynamic of romance privileges closure and embraces the promise of futurity implicit in such an ending (de Lauretis 1984; DuPlessis 1985; Roof 1996; Gill and Herdieckerhoff 2006; Roof 2007). This does not mean that there is no space for feminist rewriting in romance (Modleski 1984; Radway 1991; Homans 1994; Farwell 1996; Felski 2003; Lanser 2009). Nevertheless, the sense of closure that accompanies romantic narratives often reinforces existing gendered relations. For example, Judith Roof (1996) argues that narrative closure tends to assume that gender differences are complementary, so that the continuation of narrative depends on the coming together of oppositional elements. For Roof, narrative is characterized by an "impetus to negotiate disparate elements through the sexualized terms by which production and reproduction are conceived in Western culture" (Roof 1996: xxxii). This metaphoric connection between heterosexual coupling and narrative logic culminates in romantic narrative's literal conjoiner of two desiring parties. Romantic narrative's striving toward a union, then, tends to yield a sense of continuity and futurity that other narrative forms may suggest but rarely enforce with similar insistence.

When evoked in evolutionary discourse, the romance plot may help guarantee symbolic reproduction amidst the threat of reproductive failure. The narrative pull produced by Lodge's appropriation of the narrative dynamic of romance camouflages any doubt about the infidelity narrative's ability to produce more narrative. The familiarity of such narrative elements as courtship, recognition, crisis, and declaration of desire engenders a sense of narrative movement that implies continuity. Through this striving to futurity, the evolutionary infidelity narrative responds to what Stephanie Turner describes in the context of popular extinction narratives as the need to do "anything to keep the story going" even when "our non-existence as a species is inevitable" (Turner 2007: 76). When invoked in evolutionary psychological discourse, the romance plot often produces a self-containing loop. The productive logic of the romance plot supports the reproduction of a cultural order, which is imagined as a strictly gendered economy of prehistoric adulterous desires. These evolutionary desires, in turn, signify the promise of more courtship and hence narrative continuity.

This does not mean that the romance plot fully suppresses the threat of reproductive failure. No narrative structure holds such power. Nor does the absence of infertility imagery directly follow from the presence of the romance plot, as the presence of a textual feature is always a decision made, consciously or unconsciously, by the author. The narrative dynamic of romance does, however, produce a feeling of narrative momentum that coincides with cultural ideas of how stories should go. Such momentum seems to render insignificant a text's refusal to reward adulterous behavior with reproductive success contrary to the evolutionary psychological claims the text may invoke. In Lodge's novel, for example, there is a significant gap between the discourse of reproductive fitness that Ralph evokes and the fact that the novel's adulterous relationships seem to take place in a curiously post-reproductive cultural environment. Yet the novel's appropriation of the romance plot embeds its events in a symbolically reproductive narrative dynamic. As a result, what could be described in evolutionary terms as reproductive failure seems hardly like a failure.

Narrative mutations

I began this chapter with the suggestion that the emergence of sociobiology in the 1970s engendered a new kind of naturalization of human promiscuity by imagining reproduction as a calculative endeavor acted out at the level of gametes and genes. Premised on the antagonistic logic of warfare between the genders, sociobiology reinterpreted evolutionary transformation as driven by reproductive ambitions. While this theoretical and discursive shift evoked Darwin's idea of sexual selection by placing emphasis on reproductive success rather than survival, it departed from Darwin by imagining sexual choices as largely fixed. Evolutionary psychological texts often take these assumptions about sexuality still further, providing full-fledged narratives of genetically driven promiscuity and adultery that resonate with cultural assumptions about male and female natures. In both its androcentric and popular feminist forms that we encountered, respectively, in Ridley's and Judson's texts, the evolutionary infidelity narrative enacts in an extreme form the underlying reproductive logic of the larger evolutionary narrative. Focusing on unlimited, reproductively motivated promiscuity, the infidelity narrative sheds light on the narrative assumptions implicit in the evolutionary narrative in general.

We have also seen literary works evoke the evolutionary infidelity narrative in their vocabularies, creating characters that repeat evolutionary psychological axioms, like Gideon in Davidson's *Heredity*, Fran in Anderson's *Darwin's Wink*, or Ralph, Carrie, and to some extent Helen in Lodge's *Thinks ...* . The evolutionary infidelity narrative is also visible in the novels' framing of fundamental questions about human existence. The texts ask, for instance, whether our choices are actually ours, and what happens to morality if immoral behaviors arise from our prehistoric human natures. Most importantly, this chapter has demonstrated that a text's attitude toward evolutionary psychological discourse often takes shape through the narrative structures it deploys. Thus Lodge's appropriation and Davidson's rejection of the romance plot align the two texts, respectively, with endorsement and critique of evolutionary psychology, while Anderson's narrative politics leave her text ambivalent as to the validity of evolutionary psychological claims about human promiscuity. Read side by side, the three texts also point to an inherent structural weakness in the evolutionary infidelity narrative by underscoring the fact that the story can continue only if reproduction succeeds. In radically different ways, then, the three texts indicate how the romance plot may function as a means to symbolic reproduction and thus as a claim to futurity.

The simultaneous rejection of romantic discourse and appropriation of the romance plot in popular evolutionary discourse have manifold and complex implications. The prominence of the evolutionary infidelity narrative in the three novels attests to the extent to which sociobiological and evolutionary psychological understanding of promiscuity as a genetically coded trait has refashioned popular discourses on romantic relationships. At the same time, the three novels suggest that representations of the clash between romantic beliefs and the evolutionary psychological rejection of those beliefs are far from monolithic and unambiguous. Most importantly, the differences between the novels imply that there is space in evolutionary accounts of infidelity for narrative contestation. Such critique may take place through a revelation of the inadequacy of the narrative dynamic of the evolutionary infidelity narrative, as in Anderson's *Darwin's Wink*, or through a subtle dismantling of both the infidelity and romantic narratives, as in Davidson's *Heredity*. While the romance plot may serve a strategic purpose in evolutionary accounts of infidelity, as in Lodge's *Thinks …* , evolutionary psychological discourse can hardly contain the narrative structures it appropriates. Romance, it seems, always exceeds any particular evocation of its central narrative dynamic.

In popular evolutionary discourse, infidelity is often represented as the ultimate form of reproductive inclinations. Such extremity, however, has its polar opposite in the narrative logic of evolution. If infidelity is understood as the ultimate maximization of one's reproductive fitness, then sexual practices and desires that do not result in reproduction run against this narrative logic. The next chapter examines the many forms of exclusion that the reproductive imperative as a narrative engine generates. In particular, it explores the reliance of evolutionary psychological narratives on the understanding of narrative as a series of reproductive events. The chapter argues that such an event-based narrative dynamic is unable to account for anything that falls outside the logic of deeds, and that these exclusions eventually haunt the very logic of the adaptationist evolutionary narrative favored by evolutionary psychologists. In other words, it is precisely their claims of narrative futurity that render adaptationist narratives both most resonant with cultural discourses and structurally most vulnerable.

Notes

1 For a discussion of the theoretical continuities between Darwin and Bateman, on the one hand, and Bateman and later biologists (including Trivers), on the other, see Dewsbury 2005.
2 Moore and Desmond (2004) connect this view of racial differences to Darwin's abolitionist sympathies.
3 This line of argument has been developed by feminist sociobiologists, most famously Sarah Blaffer Hrdy (1981).
4 Roof (1996) identifies a similar logic in Freud's writings about the development of sexual identity. For Freud, the risk of perversion (homosexuality) functions as the narrative impediment that provides the development of heterosexual identity with the narrative tension, climax, and denouement that are considered intrinsic to a good story (Roof 1996: xviii–xxii). I return to this point in Chapter 5.
5 Infidelity is, of course, only one explanation that has been given to such key human developments as the birth of intelligence, language, or particular social structures. Other explanations refer, for example, to the introduction of hunting (or in some feminist versions, gathering) as a key event that led to the evolution of intelligence and language. However, the productive logic behind these alternative models is the same. They are also based on the same reproductive imperative that understands reproduction as the only means to posterity. In a sense, the evolutionary infidelity narrative is an extreme version of this productive narrative logic, since it posits the maximization of reproductive success as the ultimate goal of all action.
6 This is not unusual in evolutionary psychology, which typically argues that our traits and behaviors date back to the Pleistocene era and therefore might not be adaptive any longer.

5 Reproductive failure and narrative continuity

In 2001, the interdisciplinary literary journal *SubStance* published a special issue on evolution and imagination in which a number of literary scholars and scientists explored the evolutionary roots of art and the methodologies appropriate for its study. In an article entitled "Humanists, Scientists, and the Cultural Surplus," the issue's editor, H. Porter Abbott (2001), reflected on some scientists' accusations that scholars in the humanities refuse to take seriously evolutionary and cognitive approaches to literature. According to Abbott, such a refusal is not a matter of actual hostility toward science but rather arises from the recognition of the irreducibility of cultural phenomena:

> [W]here Wilson's kind of scientist goes wrong is in failing to understand the degree to which the humanist researcher, working in the present, must accept and work with knowledge that is, and most likely will remain, speculative. It is the willingness to think analytically in a realm in which the operating assumptions and the conclusions arrived at are not empirically grounded that may be what, at bottom, separates the humanist researcher from most scientists.
>
> (Abbott 2001: 210)

Culture, Abbott suggests, is too complex, subtle, and evasive to be reduced to a single level of explanation. Instead, "much of culture is devoted to a surplus beyond the requirements of survival" that "is no less important for that" (Abbott 2001: 215). This cultural surplus simply escapes the adaptationist logic which assumes that what exists must exist because it has been beneficial for an organism's survival and reproductive success. For literary scholars, "good interpretive arguments do not displace other good arguments that oppose them ... because the complexity of the subject does not admit of final reductions, not now at least, and not for a long time to come" (Abbott 2001: 217).

While Abbott is talking about methods of reading literature, his observation applies to other cultural phenomena as well. I want to suggest that gender and sexuality are among these phenomena. Although inseparable from the biological issues of reproduction and genetic inheritance, gender and sexuality are also products of—and productive of—a considerable amount of Abbott's cultural surplus. As such, gender and sexuality always exceed the nature/nurture dichotomy, adding up to more than the sum total of the two. Building on Abbott's idea of cultural surplus, this chapter begins with the premise that gender and sexuality are too complex to be explained exclusively through the adaptationist logic. Adaptationism is the model that constitutes the core of modern evolutionary theory. Its main premise is that inherited features arise from selective processes that favor characteristics that increase an organism's reproductive fitness—that is, the likelihood of passing its genes to a large number of offspring. Evolutionary psychology extends this adaptationist premise to all behavioral characteristics as well as many cultural phenomena, which thus appear as expressions of prehistoric reproductive ambitions. There are, of course, many biologists who do not subscribe to such extreme adaptationism (see, for example, Eldredge and Gould 1972; Gould and Lewontin 1979; Eldredge 1985; Margulis and Sagan [1986] 1997; Margulis 1998; Eldredge 2004). It is, however, the reproductively motivated adaptationist version of the evolutionary narrative that has monopolized popular discourses of science and sexuality. Furthermore, the same reproductive logic also underlies a number of evolutionary accounts that emphasize complexity and excess, albeit as an abstract structural pattern rather than as explicit theme. For example, many of the Epic of Evolution texts discussed in Chapter 2 subscribe to a logic of progress that resembles the adaptationist pattern of change.

This chapter focuses on the extreme form of adaptationism that characterizes evolutionary psychology. This kind of adaptationism organizes, for instance, David M. Buss' *The Evolution of Desire* ([1994] 2003). In this pioneering work of evolutionary psychology, Buss explains that "adaptations are human solutions to the problems of existence posed by the hostile forces of nature ... Those who failed to develop appropriate characteristics failed to survive" (Buss [1994] 2003: 5). As we saw in the Introduction, Buss casts this logic in explicitly sexual terms, as he declares:

> Those in our evolutionary past who failed to mate successfully failed to become our ancestors. All of us descend from a long and unbroken line of ancestors who competed successfully for desirable mates ... We carry in us the sexual legacy of those success stories.
>
> (Buss [1994] 2003: 5–6)

The viewpoint Buss advocates is that of the survivors: there are two kinds of organisms, those who become ancestors and those who fail to do so. At the same time, knowledge about desire and sexual behavior is assumed to be located at the very beginning of the evolutionary narrative leading to these survivors, so that "[d]iscovering the evolutionary roots of women's desires requires going far back in time, before humans evolved as a species, before primates emerged from their mammalian ancestors, back to the origins of sexual reproduction itself" (Buss [1994] 2003: 19). In such an adaptationist framework, desire appears as an object that can be isolated and identified as the monolithic engine of the events that constitute the survivors' "success stories."

In this reproductively motivated narrative of evolutionary inheritance, sex is, to quote Matt Ridley once again, the only "currency" that "counts" (Ridley [1993] 2003: 243). The kind of sex that counts, however, is not just any sex. When Buss evokes our non-ancestors' failure to mate, he is not talking about a failure to desire, to experience pleasure, or to engage in sexual encounters with others but to pass one's genes to subsequent generations. For Buss, that is, evolution is a series of reproductive sex acts. In her analysis of scientific studies of homosexual animal behavior, cultural studies scholar Jennifer Terry (2000) demonstrates that such highlighting of reproduction is not a matter of popular science distorting the discoveries of professional science, as sometimes is assumed. Rather, "[e]ven as sexual variance among animals and humans is acknowledged, most scientists continue to conceptualize sexuality narrowly in terms of the evolutionary imperative of reproduction," viewing nonreproductive behavior as something that "thwarts, disturbs, or, in the best light, merely supplements heterosexual reproduction" (Terry 2000: 154). While this heterocentrism arises from and resonates with cultural assumptions about "normal" sex and "natural" desire, it is also reinforced by the underlying logic of the evolutionary narrative. Since evolutionary narratives rely on reproductive success, as I argued in the previous chapter, what is privileged in evolutionary accounts is not heterosexuality as such but distinctly reproductive heterosexual acts. That this is indeed a structural phenomenon intrinsic to the evolutionary narrative is suggested by the fact that evolutionary accounts tend to lump together all behaviors that do not have an obvious reproductive function. As a result, a wide range of sexual behaviors—including a host of heterosexual practices—appear as evolutionary puzzles that keep scientists busy.

This chapter sets out to examine the role of nonreproductive sexuality in evolutionary narratives. In the past decade or so, a number of scholars have addressed heteronormative and androcentric assumptions in the design of tests and interpretation of their results in scientific studies of nonreproductive sexuality (Fausto-Sterling 2000; Terry 2000; Lancaster 2003; Roughgarden 2004). These critiques have often viewed extreme adaptationist arguments as symptomatic of cultural nostalgia for fixed identities (of the "boys will be boys" type) and stable gender relations (breadwinner versus homemaker). Instead of such blatant intrusions of ideology in scientific practice, this chapter explores the narrative logic that first produces nonreproductive sexual acts, pleasures, and desires as a logical impossibility and then proceeds to integrate them. The chapter also identifies a set of structural contradictions that implicitly undermine the reproductive narrative logic's privileging of adaptation over blind mutation, chance occurrence, and the idea of cultural surplus imagined by Abbott.

The chapter begins with a re-evaluation of the role of variance in Darwin's theory and its historical relationship to the evolutionary psychological understanding of human sexuality. This is followed by a discussion of three phenomena that challenge the adaptationist logic: the reproductive ambiguity of the female orgasm as a scientific object, the narrative role of homosexuality in adaptationist narratives, and the persistence of nonreproduction in Jeffrey Eugenides' novel *Middlesex*. These examples reveal conceptual exclusions that the event-based adaptationist narrative logic necessitates. They also shed light on the popular appeal of the adaptationist evolutionary narrative and the cultural anxieties it needs to negotiate. The final part of the chapter explores a gap between discourses of events and identity in popular talk on evolution. This discursive break suggests that there are limits to what evolutionary narratives may invoke, imagine, or promise in popular cultural encounters.

Mutable desires

The previous chapter argued that the narrative dynamic of the evolutionary infidelity narrative tends to rely on the fetishized act of reproductive coupling. For example, Ridley's *The Red Queen* portrays reproductive ambitions as the driving force behind what is seen as the never-ending, opportunistic conflict between men and women. Similarly, Sykes' *Adam's Curse*, discussed in Chapter 3, imagines descent as an unbroken historical line connecting the carriers of a particular Y-chromosome, whose success—as in the case of Genghis Khan—arose from the desire to maximize one's reproductive fitness. These assumptions about desire as the driving force of evolution are predicated, to quote Roger Lancaster, on the assumption that "the very vortex of origins" in distant prehistory was also "the inner sanctum of heterosexuality" (Lancaster 2003: 49). In such discourse, origins and desire are seen as defined through one another, so that desire both emerges from that origin and organizes it. Reducible to each other, origins and desire constitute a monolithic, unified entity that has no room for plurality.

In his reading of Darwin and evolutionary psychology, Roger Lancaster (2003) maintains that Darwin, too, glorifies reproductive heterosexuality. Darwin's theory of sexual selection, Lancaster suggests, undermines the progressive potential of his idea of natural selection, as his "claims about the general nature of male and female creatures rely on and require the suppression of nonreproductive, nonheterosexual sex acts from the space of nature" (Lancaster 2003: 89). While Lancaster is correct about the explicit heteronormative assumptions made by Darwin, the underlying logic of Darwin's evolutionary narrative facilitates a more ambivalent reading of the evolutionary role of nonreproductive sexualities. In other words, while Darwin endorses the gendered and heterosexist norms of his Victorian society, the logic of natural and sexual selection he proposes runs counter to such assumptions.

We saw in Chapter 1 that Darwin's theory assumes that there is no singular point of origins in the evolutionary narrative. Since change is premised on the presence of variance, among which the processes of selection take place, origins are never fully unified. As biologists Anne Fausto-Sterling, Patricia Adair Gowaty, and Marlene Zuk argue in their joint critique of evolutionary psychology, Darwin was well aware of this necessity and therefore "reveled in the varieties of life" in order to demonstrate the basis for natural selection (Fausto-Sterling *et al.* 1997: 407). According to Darwin, there is "an almost indefinite amount of fluctuating variability, by which the whole organisation is rendered in some degree plastic" (Darwin [1879] 2004: 48–9). Highlighting this role of variety in his theory, Elizabeth Grosz in fact considers Darwin as "the first theorist of becoming and the first major theorist of differentiation" (Grosz 2007: 248). For Darwin, the ongoing production of difference constitutes a powerful force in evolution, with the consequence that there is no narrative without difference. This stands in striking contrast to sociobiological and evolutionary psychological narratives, in which variance is understood as alien to the reproductive logic of evolution, and therefore as momentary excess that is quickly eliminated.

I have so far examined the connection between difference and the concept of species in *The Origin* (Chapter 1), and difference and gender characteristics in *The Descent* (Chapter 4). I shall turn now to the relationship between difference and desire in Darwin's theory. While highlighting the role of variance in Darwin's understanding of change, Elizabeth Grosz maintains that "[w]hat dictates these variations is both unknown and in some sense irrelevant, at least as far as natural selection is concerned, for it works only on the viable and inherited results of such randomness" (Grosz 2005: 26). According to Grosz:

> The range and scope of diversity and variability cannot be determined in advance, but it is significant that there are inherent, if unknown, limits to the tolerable, that is to say, sustainable variation: "monstrosities," teratological variations, may be regularly produced, but only those that remain both viable and reproductively successful, and only those that attain some evolutionary advantage, either directly or indirectly, help induce this proliferation.
>
> (Grosz 2005: 20)

In other words, Grosz considers variation as a fundamental part of evolutionary processes that constitutes both the organizing premise and ultimate outcome of evolution. Difference is, however, tolerated only insofar as it supports reproduction. Reproductive success, then, sets the ultimate parameters within which evolutionary change takes place.

While Grosz correctly emphasizes the importance of difference in Darwinian evolution, the question of the variation of desire in Darwin's evolutionary narrative is more complicated than this.[1] The previous chapter suggested that Darwin's emphasis on choice in sexual selection implicitly undermines the immutability of gender differences as narrative outcomes. This argument can be extended to sexual variance. Put simply, if choices made by organisms change, so do the desires that motivate those choices. Desire is fundamentally mutable, for unpredictable, constantly shifting traits would be unlikely to arise from a unified, immutable entity.

While Darwin seldom explicitly problematizes the idea of desire as stable and unified, his narrative politics implicitly questions this assumption. First, Darwin argues that

> as a general rule, the more diversified in structure the descendants from any one species can be rendered, the more places they will be enabled to seize on, and the more their modified progeny will be increased. In our diagram the line of succession is broken at regular intervals by small numbered letters marking the successive forms which have become sufficiently distinct to be recorded as varieties. But these breaks are imaginary, and might have been inserted anywhere, after intervals long enough to have allowed the accumulation of a considerable amount of divergent variation.
>
> (Darwin [1859] 1985: 163)

There is no clear line between norm and variation. Instead, variation is a matter of subtle and gradual changes that cannot be pinned down to any specific historical moment or locale. Second, while this variance characterizes evolutionary processes, Darwin also suggests that "the tendency to variability is in itself hereditary" (Darwin [1859] 1985: 162). What seems like unfeasible variation from the viewpoint of the reproductive imperative turns out to be something considerably more persistent, unruly, and necessary. Even though traits that dramatically compromise reproduction are not passed on to the subsequent generation, the tendency to produce ever further varieties—including new "unfeasible" varieties—is passed on through the reproducing organisms within the species. The tendency to vary, then, appears as curiously independent from the possible deviance demonstrated by the organism. Deviant and changing sexualities are neither excess necessary for the working of evolution nor the surplus of selective processes. Instead, they have a constitutive role in the narrative logic of evolutionary change. It is not just variance as such but the inherited *tendency* to variability—including sexual variability—that acts as the precondition of evolutionary change.

This logic becomes evident in the case of racial differences. If racial differences are products of individual choice and the processes of sexual selection, as Darwin suggests in *The Descent*, then races should also embody different sexualities. While Darwin emphasizes differences in physiognomy among ethnic groups, the logic of sexual selection implies that these differences should be more than skin deep. If strategies of mate choice are in constant flux, then not only external, ethnically coded appearance and behaviors but also the preference for these particular sets of characteristics is mutable. Crucially, the logic of natural selection rejects the dichotomy of external (appearance)/internal (behavior). Since there is no outside force (such as a deity) that directs evolution but instead the forces of change arise from within nature, there is also no source of desire that would be positioned outside the realm of nature. Within this framework, sexuality is only conceivable as constantly changing, because positing desire as an unchanging force would contradict the deconstruction of the nature/culture dichotomy so central to Darwin's narrative project.

This is not to say that Darwin was conscious of such far-reaching metaphysical implications of his theory. The mutability of desire is, however, what one arrives at if Darwin's theory of change is extended to its logical conclusion. Darwin himself undermined these fundamental implications by imagining the genders as expressions of a dichotomous set of gendered proclivities, which had long ago departed from the evolutionary narrative that had produced them. Indeed, Darwin concludes that

> whatever influence sexual selection may have had in producing the differences between the races of man, and between man and the higher Quadrumana, this influence would have been more powerful at a remote period than at the present day, though probably not yet wholly lost.
>
> (Darwin [1879] 2004: 663)

This idea of prehistory as the scene of key human adaptations and the present as a reflection of that history anticipates sociobiological and evolutionary psychological visions of modern humanity as driven by their Pleistoscene desires. At the same time, the very contradiction between Darwin's explicit assumptions about gender and sexuality and the logic of change he imagines attests to the ambiguity inherent in the idea of evolution. In this ambiguity lies the possibility of telling the evolutionary narrative in radically different ways. It is significant that evolutionary psychological texts choose not to embrace this multiplicity. Where the emphasis falls, then, may change the whole narrative.

Puzzling pleasures

Although variations of desire play a constitutive role in the narrative dynamic of Darwin's theory of evolution, they are commonly represented as a logical challenge in contemporary professional and popular accounts of evolution. These accounts repeatedly insist that only reproductively beneficial variation can be passed on, and that nonreproductive sexual phenomena should exist only as temporary anomalies. I shall turn now to this event-based logic of evolutionary continuity in order to examine the omissions and confusion the adaptationist narrative generates.

Female orgasm is illustrative of the adaptationist narrative logic because it occupies an ambiguous position between the realms of reproductive and nonreproductive pleasures: it neither necessarily excludes nor necessarily includes a reproductive act. As literary scholar James A. Steintrager points out, "unlike the male orgasm, which links the lure of pleasure to the survival of the species in a seemingly unproblematic fashion, the function of female orgasm is not so obvious" (Steintrager 1999: 26).[2] That female orgasm is understood as a narrative challenge is suggested by the sheer number of adaptationist accounts offered by scientists—Elisabeth A. Lloyd (2005) counts 19—which all try to incorporate female orgasm into the larger evolutionary narrative. My intention here is not to evaluate the plausibility of any particular theory, but to provide a brief overview of some of the most popular explanations of female orgasm in order to demonstrate the narrative trouble that ensues when potentially nonreproductive behaviors are embedded within an adaptationist framework.

As Lloyd painstakingly documents in *The Case of the Female Orgasm*, the vast majority of scientific explanations of female orgasm account for it as a (once) useful trait that got selected at some point in our evolutionary history. A classic among these explanations is the pair-bond account, most famously advocated by Desmond Morris (1967). Morris' account of female orgasm reflects the popular "Man the Hunter" theories of the 1960s that positioned the invention of hunting as a key evolutionary event that led to the development of social skills and thus to the emergence of modern societies. Hunting was also typically credited as having introduced the gendered division of labor, thereby placing men as the proto-breadwinners and women as their dependent partners, an arrangement understood to encourage monogamy. Most versions of the pair-bond account assume that female orgasm appeared with the monogamous family unit, encouraging women to commit themselves to only one male by creating an emotional bond (Lloyd 2005: 44–76).

Lloyd notes that scientists have also sought strictly physiological explanations. For example, the so-called "antigravity" hypothesis suggests that female orgasm encourages a woman to stay horizontal after intercourse and thus minimizes the amount of sperm escaping from the vagina (Lloyd 2005: 57–60). The so-called "upsuck" hypothesis, on the other hand, posits that the vaginal cramps during an orgasm help a woman "suck" the sperm toward the ovaries, thereby increasing fertility (Lloyd 2005: 179–219). Still others have combined physiological and psychological explanations. Frank Beach, for example, has suggested a somewhat complicated chain of events, leading from the introduction of bipedalism to the subsequent repositioning of the vagina further to the front, to the invention of face-to-face intercourse, to increased stimulation of the clitoris, and thus finally to the emergence of female orgasm (Lloyd 2005: 67–70). In this scenario, female orgasm appears as a kind of evolutionary accident.

Many of these accounts show a clear androcentric bias, such as the assumption that women trade sex for emotional fulfillment. However, it should be noted that feminist adaptationist accounts of female orgasm are not structurally different from male models. For example, Sarah Blaffer Hrdy's (1981) famous suggestion in *The Woman that Never Evolved* that female orgasm did not encourage monogamous behavior but in fact rewarded females for actively seeking multiple partners (more sex, more pleasure) reproduces the same narrative logic that motivates the theories she critiques. In Hrdy's account, females' search for orgasmic pleasure serves reproductive ends as it helps confuse paternity and thus reduce the risk of infanticide by competing males (Hrdy 1981: 174). While debunking stereotypes of women as coy and passive, feminist sociobiology yet perpetuates, to quote anthropologist Susan Sperling, "the narrative logic of functionalist models of primate behavior" (Sperling 1991: 26) that "collapse variation into theories of male and female reproductive strategies" (Sperling 1991: 27). This suggests that a critical feminist theory of evolution needs to challenge the epistemic monopoly of the reproductive imperative in order to fully account for variance in female experience and the fickleness of pleasures like female orgasm.[3]

Theories of female orgasm are illustrative of the organizing logic of adaptationist narratives also in terms of what they exclude. As many feminist commentators have pointed out, adaptationist accounts of female orgasm usually only consider so-called copulatory orgasms, reinforcing the cultural understanding of penis-in-vagina intercourse as the ultimate sex act.[4] Significantly, this focus on orgasm during vaginal intercourse is supported by the underlying logic of the adaptationist evolutionary narrative, which understands acts connected with reproduction as the only meaningful narrative events. This renders a multitude of orgasmic pleasures simply illegible. The reproductive narrative logic also interprets female orgasm as a parallel phenomenon to male orgasm. Since male orgasm is symbolically linked with reproductive success, as the event of ejaculation stands for the event of conception, female experience becomes an imperfect version of male experience. However, as Marlene Zuk points out, "[i]f we keep assuming that females, including variations among them, are not the norm, it will be hard to conclude that their responses are adaptations" (Zuk 2002: 146). Furthermore, as Nancy Tuana (2004: 221–2) and Elisabeth Lloyd (2005: 226–7) point out, some adaptationist accounts—most famously the pair-bond account—mistakenly assume that female orgasm is a uniquely human invention. By connecting the emergence of female orgasm to the human invention of the family unit, this vision of recent emergence reinforces the stereotype of women (as opposed to other female primates) as monogamous while implying that extramarital orgasms among women are in fact anomalous and thus "unnatural." This model also helps minimize the narrative trouble caused by nonreproductive sexuality: the later female orgasm arose in the human evolution, the lesser its threat to the reproductive logic of the evolutionary narrative.

Adaptationist explanations of sexuality also confuse reproductive behavior and desire. Unlike feminist and queer scholars, sociobiologists and evolutionary psychologists do not understand desire as a historically contingent, formative force that may change the outcome of the entire story. Instead, sociobiologists and evolutionary psychologists infer desire from the events that constitute the evolutionary narrative. Since reproductive acts are the only acts that count as proper events in adaptationist narratives, the desire that emerges as a result of this extrapolation is also reproductively motivated. This narrative logic finds its culmination in Dawkins' infamous selfish gene, the fantastically anthropomorphic actor that, as we saw in Chapter 3, "leaps from body to body down the generations, manipulating body after body in its own way and for its own ends," and to which organisms are mere survival machines (Dawkins [1976] 1999: 34). In Dawkins' molecular universe, the assumption that certain behaviors increase an organism's reproductive fitness becomes translated into the immutable desire of the gene. Adaptationist accounts of female orgasm further complicate this slippage between observable acts and the fickle experience of desire. Whether psychological or physiological, these accounts tend to rely on a three-step chain of inference: from behavior to pleasure to desire. Since behavior is defined as reproductive at the outset—that is, only penis-in-vagina sex counts as a narrative event—pleasure becomes equated with orgasm during heterosexual intercourse and desire with the genetically hard-wired urge to maximize one's reproductive success.

As with Dawkins' reproductively obsessed gene, popular science texts on female orgasm often portray this inferred desire as the originating actor that motivates both pleasure and the reproductive act itself. This is the case, for example, with Geoffrey Miller's evolutionary psychological exploration of human sexuality in *The Mating Mind* ([2000] 2001). Miller's text offers a three-page discussion of female orgasm in the chapter discussing "Bodies of Evidence" for sexual selection. According to Miller, the fickleness of female orgasm—the fact that penis-in-vagina intercourse does not automatically lead to orgasm—suggests that female orgasm is "a mechanism of female choice" (Miller [2000] 2001: 238) that "separates the men from the boys" (Miller [2000] 2001: 240). As Miller sees it, "[f]rom a sexual selection viewpoint, clitorises should respond only to men who demonstrate high fitness, including the physical fitness necessary for long, energetic sex, and the mental fitness necessary to understand what women want and how to deliver it" (Miller [2000] 2001: 239). Miller's scenario is circular: he starts with pleasure (the fickleness of female orgasm) and then extrapolates a reproductive strategy (the assessment of males' fitness) symbolically represented by the "choosy clitoris" (Miller [2000] 2001: 239). This reproductive strategy is understood as an articulation of desire (to reproduce). From this presumed reproductive desire, Miller again infers pleasures—orgasm in service of that desire. Crucially, this pleasure is assumed to concur with the reproductive act itself, as suggested by Miller's evocation of the "stimulatory arms race" in which "[t]he penis evolved to deliver more and more stimulation, while the clitoris evolved to demand more and more" (Miller [2000] 2001: 240). As a result, female choosiness becomes a narrative force that produces both pleasures and behaviors—but never in its own right. The desire that drives Miller's evolutionary narrative is a very particular desire deduced from the privileged act of reproductive coupling.

Miller's adaptationist account of evolution is underwritten by the accumulative logic of reproductive narrative acts. While the text represents female orgasm as a potential challenge to the coherence of evolutionary explanation, it soon integrates it within the evolutionary narrative. This takes place through the rhetoric of scientific revelation. *The Mating Mind* constructs a dichotomy between (pre-discovery) false belief and (post-discovery) true insight, thereby producing an implicit discovery narrative that operates as a claim to epistemic privilege. The way in which nonreproductive behaviors become subordinated to the reproductive logic is also indicative of the mutual embeddedness of narrative form and epistemic authority. In Miller's discussion of female orgasm, the adaptationist evolutionary narrative operates as a preset narrative framework within which behaviors become nameable, observable, and intelligible. This produces a circular argument: while the reproductive narrative logic renders adaptationism the only adequate account of nonreproductive phenomena, the very inclusion of such sexual variety buttresses the epistemic authority of the reproductive logic.

Crisis and resolution

While female orgasm may be the cause of a certain degree of narrative friction, the adaptationist narrative logic is able to explain and integrate it—even if such explanations are haunted by the myriad pleasures they exclude. But can the adaptationist logic accommodate homosexuality, a trait that is often portrayed as "the antithesis of reproductive success" (Potts and Short 1999: 74)? As scientist and science writer Robert Epstein puts it in *Scientific American Mind* magazine: "For obvious evolutionary reasons, most people are strongly inclined to prefer opposite-sex partners, because such relationships produce children who continue the human race" (Epstein 2009: 68).

One cannot blame scientists for not trying to explain homosexuality. The early 1990s in particular witnessed several attempts to address homosexuality as a biologically determined characteristic: Simon LeVay's (1991) neuroanatomical study of the gay and straight hypothalamus, J. Michael Bailey and Richard Pillard's (1991) comparative study of identical and non-identical twins and genetic and adoptive brothers, and Dean Hamer and colleagues' (1993) genetic linkage study are the most famous of these projects. Notwithstanding the enthusiasm (and horror) that surrounded the publication of their results, none of these studies addressed other than briefly and speculatively the larger evolutionary question: if homosexuality is indeed genetic, how has it survived? Yet public debates about the biological versus cultural causes of homosexuality typically portray the results of scientific studies of homosexuality as pieces in the larger evolutionary puzzle. Especially since Hamer *et al.*'s 1993 report of the discovery of a genetic linkage on the Xy28 region of the X-chromosome among gay brothers—the study that became popularized as the "gay gene discovery"—evolutionary speculations about the origins and survival of homosexuality have been a standard part of popular texts on the evolution of sexuality. As Pieter R. Adriaens and Andreas De Block observe, "it is nearly impossible to find a book on evolutionary psychology or behavioral ecology in which the author does not feel obliged at least to touch upon the theme of homosexuality" (Adriaens and De Block 2006: 571).

In popular science, homosexuality is portrayed consistently as a scientific mystery. For example, Malcolm Potts and Roger Short's *Ever Since Adam and Eve* refers to it as "a profound puzzle" (Potts and Short 1999: 74), Geoffrey Miller's *The Mating Mind* as "a genuine evolutionary enigma" (Miller [2000]

2001: 218), and David Buss' *The Evolution of Desire* as "an empirical enigma for evolutionary theory" (Buss [1994] 2003: 251). Books that focus specifically on homosexuality often acknowledge this discursive framing in their very titles, as does N. J. Peters' *Conundrum: The Evolution of Homosexuality* (2006) and Louis A. Berman's *The Puzzle: Exploring the Evolutionary Puzzle of Male Homosexuality* (2003). Like female orgasm, homosexuality is usually given an array of adaptationist explanations. These accounts have included social function theories, which often suggest that homosexuality is a way of making peace through same-sex pleasures, thus resembling the social function of homosexuality in bonobo societies. Antisocial function theories, by contrast, view homosexual acts as a mechanism of domination and subordination that produces and maintains hierarchies within the male or female community. Genetic by-product theories, on the other hand, differ from both social and antisocial theories by emphasizing invisible genetic effects. Such theories posit homosexuality as analogous to sickle cell anemia, the by-product of a resistance gene to malaria, thereby suggesting that the genetic makeup that produces homosexuality may have other clearly beneficial effects in certain circumstances.

Many of the biological explanations of homosexuality echo the gender inversion model of earlier medical literature. This model portrays gay men as effeminate and lesbians as masculine, thus confusing sexual orientation with transgender experience, as Joan Roughgarden (2004) points out. This model is manifest in early sociobiological speculations of homosexuality as a viable strategy for those (presumably effeminate) men who cannot successfully compete for mates. In *On Human Nature*, Wilson suggests that there is "a strong possibility that homosexuality is normal in a biological sense" (Wilson 1978: 143). Building on W. D. Hamilton's theory of inclusive fitness, Wilson argues that "[h]omosexuals may be the genetic carriers of some of mankind's rare altruistic impulses," which they direct toward their siblings' offspring, who share a considerable number of their genes (Wilson 1978: 143). While Wilson's argument was later largely discredited for lack of evidence, its model of gender inversion is echoed in a long line of scientific studies on homosexuality. It informs, for example, Dean Hamer's speculation on "the intriguing possibility that a genetic crossover between the male and female sex chromosomes is related to the behavioral 'crossover' between heterosexuality and homosexuality" (Hamer and Copeland 1994: 128).

As with female orgasm, it is highly unlikely that this understanding of homosexuality as an evolutionary anomaly is the result of pure objective observation by ideologically unbiased scientists. Nor is it, however, simply a matter of heteronormative ideology distorting the science. Rather, this view of homosexuality can be understood as an effect of the reproductive narrative logic that organizes evolutionary explanation. This is suggested by the fact that even accounts that do not subscribe to wholly deterministic models tend to produce a similar discursive framing. For example, while Adriaens and De Block claim that "homosexuality is not a Darwinian paradox" as evolutionary psychologists assume but "a social construction with a long evolutionary history," they assume that the emergence and persistence of homosexuality through history needs to be explained within the adaptationist framework (Adriaens and De Block 2006: 583). As a result, the "social construction" of homosexuality turns out to be yet another reproductive strategy, since it presumably allows young males to climb higher in the male hierarchy and thereby gain access to females. Similarly, an author's feminist viewpoint does not necessarily change the outcome of the story. While Olivia Judson's *Dr. Tatiana's Sex Advice to All Creation*, discussed in the previous chapter, joyfully celebrates promiscuous female sexuality, the (always adaptationist) explanations it offers for nonreproductive behaviors are almost identical to those found in mainstream popular science. This consistent framing of homosexuality as an evolutionary puzzle suggests that the reproductive narrative logic is, as Judith Roof insightfully puts it, "at the center of our understanding of evolution. For this reason, though we might assail gender, it is extremely difficult to bring our narrative about reproduction into question" (Roof 2007: 131). Heteronormativity is both evasive and all-encompassing because it operates at the level of narrative structure.

Despite its apparent challenge to evolutionary explanations, the understanding of homosexuality as a Darwinian mystery has considerable popular appeal. This appeal, too, arises from the very structures of the evolutionary narrative. First, the representation of homosexuality as a puzzle confirms the culturally marginal status of homosexuality by subordinating it to reproductive sexuality through the constant evocation of the adaptationist logic. While this is true about any kind of nonreproductive sexual behavior, the visibility and sensitivity of the question of homosexuality in contemporary culture makes this discursive marginalization particularly desirable to some audiences. This narrative framing is, of course, unlikely to persuade those readers that are firmly against both the idea of evolution and the rights of sexual minorities— a large demographic group in the United States in particular, although probably a minority among the readers of popular science texts on evolution. However, the framing is able to accommodate more general cultural anxieties about proper sexualities, especially concerning the mutability and unpredictability of sexual identities in the postmodern age. This marginalization of homosexuality is reinforced through the structural positioning of homosexual behavior in a chapter or sub-chapter toward the end of popular science books. David Buss' *The Evolution of Desire* ([1994] 2003), for example, discusses homosexuality under "Mysteries of Human Mating," a chapter added to the end of the revised edition, and Robert Wright's evolutionary psychological classic *The Moral Animal* ([1994] 2004) literally subordinates homosexuality by introducing it in an appendix entitled "Frequently Asked Questions." Similarly, it is this intrinsic subordination of homosexuality that enables Geoffrey Miller to proclaim in *The Mating Mind* that his "heterosexual emphasis comes not from homophobia, religious conviction, or moral conservatism. My subject is human evolution, and homosexual behavior is just not very important in evolution" (Miller [2000] 2001: 217).

Second, this positioning of homosexuality as an anomaly within the evolutionary narrative is attractive to many because it desexualizes homosexuality. Since the reproductive narrative logic insists that any trait that has survived needs to serve a reproductive function, homosexuality is interpreted as a reproductive strategy. Paul L. Vassey observes:

> The fitness-enhancing functions assigned to non-reproductive sexual behaviours are often seen as their primary *raison d'être*, while the sexual components of these behaviours are negated, ignored, or diminished ... Indeed, "non-reproductive sexual behavior" becomes a virtual misnomer when it is merely enacted with the ultimate purpose of increasing reproductive success.
>
> (Vassey 1998: 411)

This view of homosexual behavior as a desexualized reproductive strategy underlies, for example, Geoffrey Miller's statement that "[h]omosexual behavior—*as an adjunct to heterosexual behavior*—would be expected to evolve whenever its fitness benefits (making friends, appeasing threats, making peace after arguments) exceed its costs (energy, time, and the increased risk of sexually transmitted disease)" (Miller [2000] 2001: 218; emphasis mine). But even as a desexualized reproductive strategy, homosexuality is only the second best option. Miller continues:

> However, these [homosexual] preferences had no direct reproductive consequences, so they would have had much weaker evolutionary effects than heterosexual preferences. As a result, we have to focus on heterosexual behavior when considering the role of sexual choice in the mind's evolution.
>
> (Miller [2000] 2001: 219)

Again, it should be noted that this structural desexualization affects the representation of all nonreproductive acts or desires. However, such desexualization might be particularly desirable in the context of male homosexuality, as male homosexuality is often oversexualized in the popular imagination. Crucially, it is precisely *male* homosexuality that has been the primary object of the scientific study of homosexuality.

Third, and following from this, the adaptationist narrative logic lumps together different nonreproductive behaviors, turning them into a kind of evolutionary tool kit from which an organism chooses the most efficient strategy for its current situation. This has the effect of erasing differences among nonreproductive sexual pleasures, as such strikingly different phenomena as male homosexuality and female orgasm appear as articulations of the same reproductive imperative. Furthermore, this indifference to difference and reinterpretation of nonreproduction as reproduction also makes the reproductive imperative appear as if conclusively proven. If all seeming anomalies turn out to be reproduction in disguise, then surely reproduction is the governing principle of all phenomena. On the one hand, this allows Buss to conclude that homosexuality "seems on the surface to defy evolutionary logic" as if the whole enigmatic nature of homosexuality were merely an illusion (Buss [1994] 2003: 250).[5] On the other hand, it minimizes the need for actual scientific evidence. Peters implies this much when he states that "[f]ortunately for the purposes of this chapter, the proof of such a [gay] gene's actual existence is not required in a discussion of its evolution" (Peters 2006: 134). As with female orgasm, the all-encompassing logic of the evolutionary narrative acts as a claim to epistemic privilege and, by implication, to cultural authority.

The integration of the "puzzle" of homosexuality into adaptationist discussions of the evolution of sexuality ends up confirming the reproductive logic that nonreproductive sexuality seems to challenge. As the seemingly insolvable yet (as it turns out) solvable test case for evolutionary theory, homosexuality appears as if proving the explanatory power of the evolutionary narrative. Again, this confirmatory function is an effect of narrative dynamic. In her analysis of Freud's narrative of the development of adult sexuality, Roof observes that homosexuality functions as the "threat, risk, conflict, impediment, or motive" that makes Freud's narrative a good story (Roof 1996: xix). While homosexuality "acquires its meaning as perversion precisely from its threat to truncate the story," such "perversions are absolutely indispensable to the story; their possibility and presence complicate the narrative of sexuality, making Freud's story the right story—right because it is a narrative instead of the simplistic developmental trajectory" (Roof 1996: xxi). In evolutionary psychological accounts of sexuality, too, homosexuality provides the narrative danger that the evolutionary narrative needs to negotiate. This narrative crisis functions to complicate the otherwise straightforward plot and thereby to reinforce the symbolic significance of the ending: what appeared initially as a threat becomes domesticated under the reproductive principle that underlies both evolution and narrative. As a result, narrative movement and epistemic authority appear as fundamentally inseparable.

As we saw in the context of female orgasm, the adaptationist narrative logic tends to fetishize reproductive acts, extrapolating both pleasure and desire from those acts. Discussions of homosexuality, too, tend to conceive of homosexuality in terms of the sex acts it generates. This is partly a result of the narrative logic itself: since a narrative needs events to keep the story going, homosexuality becomes equated with the same-sex sex act. However, only (seemingly) reproductive acts count as proper narrative events, that is, as events that turn the fittest of us into future ancestors. This suggests that homosexuality should logically produce *anti-acts*, acts that threaten reproduction and thus the survival of the species, the gene, and the narrative. By portraying homosexuality as a reproductive strategy, adaptationist evolutionary narratives are able to explain these anti-events as proper narrative events. Such an obsession with acts, however, is unlikely to produce explanations of desire, pleasure, or identity as anything other than hypothetical extrapolated objects in service of these strategies. As Lancaster puts it, evolutionary psychology's "attempts at reducing desire to a thinglike object of study give the illusion of stability and permanence to relations and practices that are ambiguous, contested, and in flux" (Lancaster 2003: 20).

Furthermore, the homosexual acts that evolutionary narratives imagine are almost exclusively acts between men. To a large extent, this results from the fact that scientific inquiry and cultural practices have historically privileged male experience. As Jennifer Terry notes:

> [T]he invisibility of female homosexuality is, in significant ways, guaranteed by the assumptions and techniques of these studies. One cannot simply invert the hypotheses concerning the biology of male homosexuality and come up with an adequate biological—let alone cultural—explanation for lesbianism, even though many scientists suggest as much.
>
> (Terry 2000: 159)

At the same time, the disappearance of lesbianism may also be partly due to the fact that the very idea of a lesbian sex act has often been a conceptual paradox in Western culture. This "ghosting of the lesbian," as Terry Castle calls it, renders female homosexuality a social arrangement or identity instead of seeing it as a specific desire or carnal act (Castle 1993: 5). This status of *non-act* makes female homosexuality emerge as fuzzy and distant in evolutionary narratives, as any non-act falls, by definition, outside the scope of the adaptationist imagination. As a non-act, lesbianism also poses a lesser threat to the reproductive narrative logic than male homosexuality, the source of what is commonly imagined as highly sexual acts. It is male homosexuality, then, that adaptationist evolutionary narratives evoke, explain, and eventually domesticate.[6]

Narrative trouble

Nonreproductive sexualities, however, are more resistant than first may seem. While evolutionary narratives are able to accommodate homosexuality as a desexualized (or even heterosexualized) act in service of reproductive fitness, they are not able to dispose of the homosexual (or any other nonnormative) organism as a potential source of future narrative trouble. Most importantly, the adaptationist logic is not able to resolve the fact that reproductive sexuality may generate nonreproductive sexualities. As in previous chapters, we shall turn to a work of fiction, Jeffrey Eugenides' Pulitzer Prize-winning and widely acclaimed novel *Middlesex* (2002), in order to examine the narrative trouble engendered by the future possibility of reproductive failure. A number of scholars have addressed Eugenides' representation of intersexed embodiment as a challenge to binary thinking (Collado-Rodríguez 2006; Sifuentes 2006; Shostak 2008; Hsu 2011) as well as intersex as a historical, biological, and cultural phenomenon (Dreger 1998; Fausto-Sterling 2000; Morland 2001;

Karkazis 2008; Reis 2009). This chapter, however, focuses on the narrative of genetic transmission that Eugenides offers. Read in the context of evolutionary narratives, *Middlesex* suggests that it is precisely the lurking presence of future anti-acts that threatens the coherence of the adaptationist evolutionary narrative.

Middlesex includes two intertwined narratives of inheritance. On the one hand, the novel tells the story of immigration, assimilation, and cultural transmission in twentieth-century America as acted out by the narrator Cal Stephanides' immigrant grandparents and his all-American Midwestern parents. On the other hand, the novel depicts Cal's own experiences as a carrier of two copies of a rare recessive genetic mutation that causes 5-alpha-reductase deficiency syndrome, an intersexed condition that turns a chromosomally male (XY) embryo into a seemingly normal baby girl before birth. Born and raised as a girl, Cal(liope) is faced with unusually confused teenage years as the onset of puberty increases the levels of testosterone in his body, activating an ambiguous bodily metamorphosis and leading eventually, at the age of fourteen, to the medical elite's discovery of his intersexed genitalia. These two narratives are tightly interlocked in a causal logic, as Cal's condition is represented as the outcome of the "roller-coaster ride of a single gene through time," which in turn has been enabled by a series of incestuous or genetically homogeneous marriages (his grandparents are siblings, his parents second cousins) running back to eighteenth-century Asia Minor (Eugenides 2002: 4). As Debra Shostak puts it, "Eugenides enfolds the story of incest within the narrative of geographical migration in such a way as to imply a causal connection—as if the Stephanides' survival on the way to America required transgression against difference" (Shostak 2008: 393). For the novel's narrator, the two narratives merge under the trope of inheritance: "Some people inherit houses; others paintings or highly insured violin bows. Still others get a Japanese tansu or a famous name. I got a recessive gene on my fifth chromosome and some very rare family jewels indeed" (Eugenides 2002: 451). Like other inherited objects, Cal's genetic mutation comes to stand for immaterial forms of cultural inheritance—family traditions, food preferences, taste in music—signifying a specific ethnically, economically, and gender-coded way of life.

As many have noted, Cal's intersexed body functions as a symbol for the cultural, intergenerational, and personal conflicts and negotiations that the novel portrays. For example, Francisco Collado-Rodríguez maintains that

> the protagonist's undefined gender and sexual identities become a bodily extension of the hybrid racial condition of all her/his family—and of all immigrant families who went to the new promised land of America or who were already there when the first Anglo-Saxon colonists arrived.
>
> (Collado-Rodríguez 2006: 74)

At the same time, Cal's ambiguous embodiment functions as a symbolic space in which competing theories of gender and sexuality collide. Like the contested mutilated body in the famous John/Joan sex reassignment case that Judith Butler analyzes in *Undoing Gender*, Cal's body becomes, to quote Butler's words, "a point of reference for a narrative that is not about this body, but which seizes upon the body, as it were, in order to inaugurate a narrative that interrogates the limits of the conceivably human" (Butler 2004: 64). Cal's embodiment, that is, functions differently in the different narratives within which it becomes embedded. As such, it becomes a nexus of attention through which the limits of cultural intelligibility are negotiated in the novel.

The two theories that contest for Cal's body are social constructionism of the 1960s and 1970s, and the gene-centered sociobiology that followed shortly after. The first of these positions echoes psychologist and sexologist John Money's pioneering and controversial studies on transgender and intersex, in which he advocated the view that gender identity is a product of nurture.[7] In *Middlesex*, this position materializes in Dr Luce, the gender identity expert who examines Cal after the discovery of his ambiguous genitalia and proclaims him a girl on the basis of his rearing. The second, evolutionary psychological position takes shape through the narrator's own commentary. For example, Cal is "thinking E. O. Wilson thoughts" when contemplating his grandparents' incestuous marriage (Eugenides 2002: 42), describes his family's escape to the attic during the 1967 riots in Detroit as perhaps "a vestige of our arboreal past" (Eugenides 2002: 270), calls puberty a time when "[d]eadlines encoded in the species are met" (Eugenides 2002: 323), and speculates on his former popularity in the all-girls school as perhaps being due to his body having "released pheromones that affected my schoolmates" (Eugenides 2002: 344). At a more general level, evolution functions as a metaphor for the narrator's experiences and observations, as when he depicts the bathroom in the school basement as having "a time frame I felt much more comfortable with, not the rat race of the school upstairs but the slow, evolutionary progress of the earth, of its plant and animal life forming out of the generative, primeval mud" (Eugenides 2002: 372).

Neither the constructionists nor the sociobiologists are able to fully account for Cal's embodiment. They both fit Cal within an existing narrative of gender and sexuality, thus reducing Cal's embodiment and experience into a set of false binaries. The failure of the constructionist camp is obvious as Cal escapes the surgery prescribed by Luce and decides to live as a man. The failure of evolutionary psychological explanation is less obvious, since Cal himself repeatedly speculates (even if somewhat light-heartedly) on his "true biological nature" as having guided his behaviors and his desire for the Object, his best friend in the all-girls school he attends as a teenager (Eugenides 2002: 370). Where the evolutionary account runs short, however, is in dealing with the constant reappearance of the recessive trait in the evolutionary narrative. Crucial here is the fact that, notwithstanding his choice of gender identity, Cal will be unable to have children. Like the stubbornly nonreproducing homosexual that evolutionary narratives invoke, Cal is a living narrative paradox that, from the adaptationist point of view, really should not exist. Darwin himself suggests this much when he argues that "we may feel sure that any variation in the least degree injurious would be rigidly destroyed" (Darwin [1859] 1985: 131). Dean Hamer puts this point even more bluntly in his account of the "gay gene discovery" as he states that "[t]he cold, calculated process of evolution is mercilessly unkind to genes that don't contribute to reproduction, cleansing these genes from the species, causing them to die out quickly" (Hamer and Copeland 1994: 180). Yet the narrative politics of *Middlesex* subtly challenges this logic.

First, Eugenides makes clear that Cal's condition is not a recent one but has been reappearing for centuries in the isolated village of his grandparents' childhood. That is, the infertile mutation persists, reproducing itself against the pressures of natural selection and, most importantly, against the very logic of the evolutionary narrative in which only an ongoing chain of reproduction counts as a satisfactory narrative outcome. Second, the mutation is represented as a crooked version of the Dawkinsian selfish gene. Created by the mock-Greek "biology gods, for their own amusement" (Eugenides 2002: 238), the mischievous anthropomorphic gene obeys no one, being only interested in "[h]itching a ride" (Eugenides 2002: 238) and "ensuring its expression" (Eugenides 2002: 42). This displacement of agency onto the molecular (or divine) narrative level renders humans mere carriers, who, while reproducing like Cal's grandparents and parents, in fact breed infertility. This positioning of humans as blind victims is further emphasized by the text's repeated references to the impossibility of detecting the recessive gene with bare eyes, as when the narrator describes the medical examination awaiting immigrants on Ellis Island: "no matter how well trained, medical eyes couldn't spot a recessive mutation hiding out on a fifth chromosome. Fingers couldn't feel it. Buttonhooks couldn't bring it to light" (Eugenides 2002: 93–94). Since the recessive mutation will eventually halt reproduction, the invisibility of the mutation stands for the invisibility of the narrative threat of nonreproduction. This invocation of invisibility echoes Simon Mawer's treatment of genetic mutations in *Mendel's Dwarf.* As one of Mawer's characters notes: "Recessives, that's the name of the game. Recessives play on people's anxieties" (Mawer 1998: 68).

What *Middlesex* suggests is that distinguishing reproduction from nonreproduction is a difficult and deceptive undertaking. Early in the novel, Cal describes his own existence as an end product of his ancestors' lives:

> In any genetic history, I'm the final clause in a periodic sentence, and that sentence begins a long time ago, in another language, and you have to read it from the beginning to get to the end, which is my arrival.
>
> (Eugenides 2002: 22)

While Cal is literally the final stop in the genetic history he is part of, he is also a product of a long line of successful matings celebrated by David Buss and other evolutionary psychologists. Cal's ambiguous embodiment, then, does not threaten the social and cultural *status quo* simply by resisting categorization. As a threat produced *through* reproductive success, Cal's body challenges the reproduction/nonreproduction binary itself. Reproduction, the text suggests, may give rise to the possibility of nonreproduction by passing on genes that, from the adaptationist point of view, should not be passed. Reproduction, in other words, may hide its own antidote. At the same time, nonreproduction, as future potentiality to halt reproduction, inheres in any act of reproduction. This also means that what is cast in doubt in *Middlesex* is not just the concept of reproduction but also the very concept of narrative event. If reproduction, the true narrative event, turns out to be nonreproduction in disguise, then any evolutionary event carries the seed of an evolutionary anti-event.

This concern about "passing" gains further resonance from the fact that Cal is involved in several nonreproductive couplings that blur the border between reproduction and nonreproduction—first as a seemingly heterosexual girl with the Object's brother, then as a closeted lesbian with the Object, and finally as a symbolically fertile but in fact infertile middle-aged man. In all these relationships, he, like his genes, is never quite what he seems and is repeatedly mistaken for what he is not. Even after over two decades as a man, he is not quite a guy but rather "like a guy" (Eugenides 2002: 122). This passing is, first and foremost, symbolic: Cal's ambiguous gender (and thus sexuality) suggests a transgression against the reproductive assumption underlying conventional masculinity and femininity. In this sense, one form of passing (that of gender identity) resonates with another (that of genes), pointing to the social norm that insists that masquerade is always fraud. Furthermore, while Cal's nonreproductive sexual encounters are evolutionary anti-acts, they differ from the anti-acts male homosexuality is imagined to generate. Crucially, Cal's acts involve people with (if not real, at least symbolic) reproductive potential, (unintentionally) confusing and misleading them. Many of his sexual relationships are understood as dangerous because they are anti-events passing as events, nonreproduction in the guise of potential reproduction. The reproductive imperative that evolutionary psychologists in particular like to imagine as guiding our desires, then, turns out to generate much more than just reproduction. It generates hidden infertility, unexpected narrative trouble, and cultural anxieties about how to contain the two. Most of all, it reproduces the possibility of acts that may cause the adaptationist narrative logic to collapse.

Identity crisis

This chapter opened with the suggestion that the reproductive dynamic of evolutionary narratives is unable to fully account for what Abbott calls "cultural surplus." As we saw in the case of the female orgasm, the adaptationist narrative logic privileges clearly defined reproductive acts over pleasures and desires, thereby excluding a whole field of knowledge and experience. However, there is another crucial disparity between the adaptationist narrative logic and the cultural imagination. This is the contrast between the event-based evolutionary logic and the widely endorsed idea of sexuality as an integral part of a person's identity. It is this tendency to understand sexuality in terms of identities that characterizes much of today's popular talk on homo-sexuality. The rest of this chapter addresses the question of cultural surplus in the context of identity discourse in popular representations of science and sexuality. It situates the evolutionary narrative within the public debates about the validity and implications of research on homosexuality in the past two decades. Through this shift of focus, I explore the conceptual friction between narrative structure and cultural context that underlies the collision between the event-based logic of evolution and the identity-centered discourses of sexuality.

As we have seen, evolutionary accounts understand narrative as a sequence of events that is productive of more narrative. In adaptationist narratives in particular, the production of more narrative and the reproduction of traits become symbolically equated. As a result, a narrative event is assumed to imply reproductive ambition, which in the retrospective framework of evolution is often translated into reproductive success. While this logic turns non-reproductive phenomena into non-acts, it also runs counter to the idea of identity as the seat of agency. If evolution is driven by the genetically coded urge to reproduce, then identity turns out to be a mere illusion. At the same time, this very illusion of the realness of identity is understood as the end-product of evolutionary processes. As with desires and pleasures that escape the reproductive narrative logic, identities, too, become intelligible only as inferred, reduced, and trivialized side-effects of the all-encompassing narrative of evolutionary adaption.

In popular discourse, homosexuality is imagined predominantly as a question of identity. Rather than another reproductive strategy, the homosexuality of public defenses of same-sex rights appears as arising from an innate and fixed subjectivity. The way in which the genetic and neuroanatomical studies on sexual orientation in the early 1990s were received in the media is indicative of this conflict between the discourses of events and identities. Sociologist Sarah A. Wilcox (2003) has analyzed science journalism on sexuality in the United States in the 1990s. She demonstrates that studies on sexual orienta-tion were typically framed within the larger debate over the innateness versus acquisition of sexuality. Wilcox argues that the discourse of inborn identities did not emerge primarily from scientific practice. Rather, the studies that received most publicity in the media did so "not only because the science pushed the question of biology into the public discussion, but because these studies could be pulled into the existing cultural discourse" (Wilcox 2003: 235).[8] Wilcox also observes that the assumptions about innate identities in media coverage often exceed genetic discourse:

> [T]he words "biological" and "genetic" were frequently used as shorthand references to the broad and complex concept of what a person is born

with. This broad concept can include the idea that being born gay is a form of self-knowledge rather than scientific knowledge and sometimes has a strong religious underpinning, in which sexuality is seen as given by God.

(Wilcox 2003: 232)

Scientific studies of sexual orientation, then, were given meaning within a discursive framework that resembled the conceptual models of the scientific endeavor only superficially. At the same time, Roger Lancaster observes, the scientifically dressed "new discourses on 'nature' expand the space of identity politics," so that the science of sexuality becomes another field in which larger cultural battles are fought (Lancaster 2003: 271).

There are key differences between British and American media responses to scientific studies of homosexuality. Sociologists Peter Conrad and Susan Markens (2001) analyze the British and American press coverage of Hamer's 1993 linkage analysis on male homosexuality and its follow-up study in 1995. Conrad and Markens note that the American reception was generally optimistic and even enthusiastic, although somewhat careful about the finality of the results. The British reception, however, had a clearly pessimistic and even alarmist tone, a difference that Conrad and Markens attribute to national tendencies in press culture, gay activism, and the general attitude toward science as a means to progress. Interestingly, Conrad and Markens also observe that the American press coverage rarely set the question of homosexuality within an explicit evolutionary framework. The British reception, by contrast, evoked the idea of homosexuality as an evolutionary puzzle. Conrad and Markens note that this produced an intriguing contradiction within the British reception, as "the British press presented Hamer's study with skepticism and fraught with peril, yet at the same time their reporting of the research amplified the idea that there actually is a 'gay gene' which could create such apprehension" (Conrad and Markens 2001: 387). By evoking the evolutionary paradigm, that is, the British press constructed the object they critiqued as a factual, physical entity—no matter that Hamer's study did not actually identify a gene for homosexuality but merely located a genetic marker on the X chromosome with hypothetical relevance for homosexuality. The gay gene appeared paradoxically both as the source of homosexual identity, which the possibility of genetic engineering and selective abortion threatened, and as a seeming evolutionary anomaly that needed explanation. This contradictory representation posits the colliding discourses of identity and evolutionary adaptationism as if belonging to different narratives of sexuality. The first of these implicit narratives traces the struggles of an individual gay subject who seeks self-expression and recognition. The second conjures a world inhabited by Dawkinsian survival machines, for whose genes survival is not a matter of individual success or tragedy but of statistical variation in the gene pool.

The accounts of male homosexuality examined earlier in this chapter confirm that this discursive confusion is not an exception in popular representations of science and sexuality. For example, *The Science of Desire* (1994), Dean Hamer's personal account of the "discovery" of the "gay gene," represents Hamer's research as an investigation into the complexities of homosexual identity.[9] Because of this focus, Hamer "wanted to know about [his] subjects' sexual thoughts, fantasies, behaviors, and attitudes and how these had unfolded over the course of their lives" rather than merely their sexual behavior (Hamer and Copeland 1994: 47). Biologist Joan Roughgarden argues that Hamer's focus on identity is methodologically problematic, as it is doubtful "whether any hypothetical gene for homosexuality would pertain more to perceived identity than to practice" (Roughgarden 2004: 250). Furthermore, while evoking the idea of identity, Hamer nevertheless interprets sexuality as a molecular detail rather than a characteristic of a person. In a seven-page chapter titled "Evolution" toward the end of the book, this molecular-level difference is further situated within an adaptationist framework, as Hamer speculates on the possible evolutionary function of homosexuality. As a result, sexuality in general and homosexuality in particular emerge as functions of the human genome, which in turn is understood as the outcome of the adaptationist principle. This is further complicated by Hamer's occasional references to "the fluid nature of sexuality" (Hamer and Copeland 1994: 54). According to Hamer, sexuality is "difficult to quantify … and make it fit into the neat charts and graphs so loved by scientists" (Hamer and Copeland 1994: 51). This invocation of fluidity and evasiveness contradicts both the identity discourse and the adaptationist logic.

N. J. Peters' discussion of homosexuality in *Conundrum* seems even more confused about these competing discourses. Peters constructs a dichotomy between human and animal sexualities, suggesting that while "an animal's desires are only inferred," the case for humans is more complicated, since "homosexual behavior is generally homoerotic as well—it involves emotions and feelings" (Peters 2006: 65). This dichotomy mistakes the fact that we cannot access animals' minds for the assertion that animal experience is categorically different from human experience. It also dismisses the fact that evolutionary explanations focus on sexuality as observable behavior rather than as personal, often fickle experience. Most importantly, by defining true homosexual desire as an exclusively human characteristic, the text contradicts the evolutionary premise that humans are ultimately animals. This portrayal of homosexuality is further complicated by the statement that "cultural variability in the expression of homosexuality … cannot be used as evidence that there is not a gay gene, in that the anthropological record is essentially silent on the question of *desire*" (Peters 2006: 137). By escaping both evolutionary and anthropological observation, homosexuality falls altogether outside the reach of science. The same does not apply to heterosexual desire. Even though adaptationist narratives imagine heterosexual desire only as extrapolated from reproductive sex acts, heterosexuality nevertheless appears as the organizing force behind the evolutionary narrative. There is a clear difference between reproductive and nonreproductive desire. While the former exists only in a drastically reduced form, the latter simply disappears.

This collision between identity discourse and adaptationist logic also underlies scientific and popular accounts of animal sexuality. Jennifer Terry (2000) argues that studies of homosexual behavior in animals often project uniquely human concepts such as "identity" or "desire" on the animal world. At the same time, such studies tend to reinforce the assumption that animal behavior can tell us something significant about complex human experiences. This two-way discursive traffic is highly problematic: "In studies of laboratory animals, the simplicity of observation and quantification of their sexual behavior disguises the problem that overt behavior alone cannot tell us much about sexual fantasy, psychically based libidinal investments, or the complexity of sexual desire" (Terry 2000: 165). This becomes evident in science writer Emily V. Driscoll's article "Bisexual Species" in a *Scientific American Mind* special issue on the biology of sexuality (Driscoll 2009). Echoing the texts discussed above, Driscoll's article posits nonreproductive sexuality as an evolutionary puzzle. According to the article, homosexual behavior may "ease social tensions" (Driscoll 2009: 23), "promote bonding" (Driscoll 2009: 23), or "maintain fecundity" (Driscoll 2009: 21), and thus "might confer an evolutionary advantage in some circumstances" (Driscoll 2009: 24). Not all these functions of homosexuality follow the strictly reproductive logic of evolutionary psychology. The function of easing social tension, for example, repeats assumptions familiar from earlier functionalist accounts of animal behavior. Nevertheless, such explanations rely on the idea that a behavior has to fit within the larger evolutionary frame in order to justify its persistent existence.

Driscoll's depiction of homosexual behavior in animals echoes the discursive confusion familiar from texts on human homosexuality. On the one hand, Driscoll maintains that, unlike human behavior, animal behavior does not follow a clear distinction between homosexuality and heterosexuality, as animals that exhibit homosexual behavior tend to also engage in heterosexual encounters. This point is underlined by a quote from sociologist Eric Anderson: "Animals don't do sexual identity. They just do sex" (Driscoll 2009: 21). On the other hand, the text constantly evokes cultural discourse on sexual identities. Driscoll's title, "Bisexual Species," uses bisexuality not as a technical term for observable sexual behavior but as an attribute that suggests a personality trait. Similarly, the text uses the word *gay* at several points: one of the sidebars is titled "Fast Facts: Fit to Be Gay" (Driscoll 2009: 22) and another "Let Them Be Gay" (Driscoll 2009: 24), and there are references to "gay behaviors" (Driscoll 2009: 22) and "gay penguin couplings" (Driscoll 2009: 24). While this use of *gay* could be explained as popularization that aims at selling science to non-specialist audiences, it is also indicative of the more fundamental collision between different conceptual frameworks. In other words, the models used by scientists to explain homosexuality may not make much sense in a cultural context infused with discourses of identity, agency, and personal fulfillment.

In the texts studied above, this discursive collision results in a weird mixture of permanent identities that escape the impact of history and a narrative of historical change that builds on a hypothetical chain of events behind those identities. This conceptual slippage surrounding homosexuality underscores the evolutionary narrative's problematic relationship to identity in general, suggesting that evolutionary narratives are not immune to conceptual challenge. The unresolved conflict between discourses of identity and the adaptationist logic, then, points to the limitations of the evolutionary narrative as the explanatory context of complex cultural phenomena.

The narrative surplus of sexuality

This chapter has argued that nonreproductive sexuality, despite its seemingly antithetical nature, becomes domesticated under the adaptationist logic in evolutionary narratives of sexuality. As a tricky yet solvable test case for the adaptationist enterprise, nonreproductive sexuality provides both the story of evolutionary progression and the story of scientific advance with the narrative crisis that makes narrative resolution possible. Rather than a structural threat to evolutionary narratives, nonreproductive sexuality turns out to be an intrinsic, even necessary part of those narratives. I have also suggested that this narrative logic has inherent weaknesses. First, the concept of a reproductive event that is so central to evolutionary narratives depends on successful reproduction. Any challenge to the distinction between reproduction and nonreproduction undermines the logic that posits reproduction as the primary impetus of evolutionary narrative. As Eugenides' *Middlesex* demonstrates, nonreproduction is a product of reproductive success, and, conversely, in reproduction inheres the very possibility of a future reproductive failure. Second, the adaptationist evolutionary narrative is able to account for sexuality only as a reproductive act. This privileging of acts effects a series of exclusions, as sexuality is reduced to its observable practices.

While the rhetorical appeal of adaptationist models arises from their seeming narrative coherence, that coherence also limits their explanatory scope. Adaptationist thinking can explain only so much, and it can do it only from a particular, situated point of view. When Hamer evokes "the fluid nature of sexuality" and states that sexuality is "difficult to quantify," the narrative framework within which he places that sexuality turns what is fluid and unquantifiable into fixed, extrapolated objects. These limitations challenge the epistemic authority of the adaptationist narrative. If pleasures, desires, and identities disappear in this narrative, then surely it cannot be the whole story. If the narrative constantly threatens to fall apart, surely it is not the final truth. Most importantly, the adaptationist evolutionary narrative relies on a denial of the cultural surplus. When surplus is recognized, it is represented not as irreducible variance but as raw material or a by-product necessary for the evolutionary process, and thus as ultimately productive of the evolutionary narrative. This dismissal of the narrative surplus of sexuality constrains evolutionary narratives' ability to explain, imply, and resonate. It is by identifying these structural weaknesses in evolutionary narratives that feminist and queer inquiries may challenge the discursive prominence of evolutionary psychology.

We have also seen that evolutionary accounts of nonreproductive sexuality are inherently intertwined with claims of epistemic privilege. That evolutionary narratives frame nonreproductive sexualities as an interpretative challenge and then proceed to explain them is indicative of the complex ways in which narrative and ideology are entangled. As foundational narratives of the human condition, evolutionary narratives assert themselves as authoritative interpretations of what counts as knowledge (epistemology) and what counts as being (ontology). Evolutionary narratives operate as interpretative frameworks within which all phenomena are labeled and evaluated. As a result, the organizing logic of evolutionary narrative affects what may count as scientific fact or valid evidence. In evolutionary accounts of nonreproductive sexualities, assumptions of adaptationism often replace empirical evidence, covering up the fact that there is very little evidence for how a hypothetical DNA sequence might translate into a highly complex trait like sexuality.

In Darwin's writing on evolution, nonreproductive sexuality appears as constitutive of evolution as an inherited potentiality which may or may not produce deviant sexualities. This potentiality to desire differently invests Darwin's evolutionary narrative with a sense of danger and unpredictability that renders the movement of evolutionary processes fundamentally precarious. While evolutionary psychological accounts understand nonreproductive sexualities as natural variance, they seldom consider them as really that different from reproduction. Rather, nonreproductive sexualities emerge as strategies that may contribute to an organism's reproductive fitness, thereby reinforcing the event-based logic of movement that posits evolution as a linear foundational trajectory. These continuities and discontinuities between Darwin, sociobiologists, and evolutionary psychologists attest that the connections among science, culture, and ideology are complicated and shifting. Most importantly, they suggest that the way in which today's adaptationists frame their stories of sexuality is not an inevitable outcome of Darwin's textual politics.

Notes

1 Both Luciana Parisi (2010) and Jami Weinstein (2010) identify a problematic preference for a dualistic system of gender and sexuality in Grosz's writing on evolution. As Weinstein points out, this view arises from Grosz's engagement with theories of sexual difference. However, Parisi argues, "[i]f evolution can explain the ontological dynamics of sexual difference, it has to be radically addressed as an incomplete process, open to the *unthought* of sex" (Parisi 2010: 155). While I agree with Parisi and Weinstein, my focus is not on Grosz and the ontology of sexual difference but Darwin and the narrative logic of evolution.

2 Some feminist biologists have challenged the assumption that male orgasm is a highly perfected evolutionary adaptation. Marlene Zuk asks: "What is so perfectly efficient about male orgasm, after all? Men, generally speaking, do not ejaculate the instant their penis is inserted; in virtually all mammals a period of multiple intromissions or thrusting is required. Why is that not inefficient?" (Zuk 2002: 145).

3 For a thorough discussion and critique of adaptationist theories of female orgasm, see Lloyd (2005).

4 For critique of the exclusion of noncopulatory orgasms, see Lloyd (2005: 223–4) or Tuana (2004: 219–22).

5 Peters employs similar rhetoric in *Conundrum* when he writes that "[i]t appears *on the surface* that there cannot be a simple gay gene" (Peters 2006: 137) and when he refers to homosexuality as "an apparent conundrum" (Peters 2006: 156).

6 This focus on reproductive sex acts in evolutionary narratives is further challenged by the fact that many lesbian and gay couples have children. See, for example, Dennis W. Allen's (1995: 617–18) discussion of the 1993 State Supreme Court of Hawaii ruling on *Baehr v. Lewin*, a lawsuit that questioned the state's rejection of same-sex marriage, and the consequent introduction of a bill that defined marriage in terms of reproductive potential. Allen points out that such a framing inadvertently questioned the coherence of the heterosexual narrative by drawing attention to the use of reproductive technologies both among infertile heterosexual couples and among lesbian and gay couples.

7 Money is also known for his involvement in the controversial "John/Joan" case. Money was the doctor who recommended that the baby boy later known as David Reimer should be raised as a girl after his penis was severely burned during a botched surgery at the age of eight months. Eugenides refers to Money's *Venuses Penuses: Sexology, Sexosophy, and Exigency Theory* (1986) in the acknowledgments at the beginning of the novel.

8 Wilcox concludes that "political debates over sexuality, with the underlying assumed dichotomy between biology and choice, are the primary frame through which scientific research studies are defined as newsworthy and receive high levels of media attention" (Wilcox 2003: 233–4).

9 The book is co-authored with journalist Peter Copeland. However, the text is written in the first person, with Hamer as the narrator.

Conclusion

This book has investigated the relationship between contemporary evolutionary discourse and earlier forms of evolutionary imagination. In particular, I have traced the transformations of evolutionary narrative from Darwin to sociobiology to evolutionary psychology. The previous chapters have portrayed this relationship as one of appropriation, revision, and reinterpretation of the narrative potential inherent in Darwin's theory. I have hoped, on the one hand, to shed light on the imaginative limits of the evolutionary narrative, outlining the structural constraints to which evolutionary narratives must conform in order to remain *evolutionary* narratives. On the other hand, I have explored the narrative potential latent in evolutionary theory.

While exploring the limits and possibilities of narrative transformation, the book has examined gender, sexuality, and reproduction at the level of narrative structure. The book opened with the observation that feminist and queer scholarship often draws connections between the epistemic endeavors of evolutionary psychology and reactionary, heterosexist, and male chauvinist politics. This book has sought to specify and complicate these claims by connecting ideology to the structural properties of evolutionary narratives. In particular, I have investigated the ways in which evolutionary narratives' assumptions of stability and movement reinforce particular politics of knowledge, authority, and cultural change. By setting the question of ideology in terms of narrative dynamic, we may understand the persistence and evasiveness of gender ideologies in contemporary evolutionary discourse.

The previous chapters have indicated that cultural context is crucial to the historical negotiations about where narratives can (and should) and where they cannot (or should not) go. It is the cultural context that sets the ultimate parameters for the evolutionary imagination. While the interface between narrative structure and cultural context is often marked by considerable friction, it does not follow the strict dichotomy of intratextual versus extratextual properties. Narrative is immersed in culture from its very conception, and the very idea of narrative structure reflects a particular way of producing and organizing knowledge. This means that there is no strictly extratextual dimension to evolutionary narrative, nor is there anything distinctly intratextual to narrative that would not already be part of a larger cultural dynamic. We saw this in the case of the reproductive narrative logic, which echoes not only explicit, culturally specific assumptions about gender and sexuality but also cultural ideas of what constitutes a proper story. Similarly, gender and sexuality are not simply sets of cultural ideas that are represented in evolutionary narratives but these ideas are implicated in the underlying structure of those narratives.

Disruptions and continuities

The clearest discontinuity between Darwin and evolutionary psychology is their attitude toward humans' self-control over sexual urges and acts. Whereas Darwin considers the conscious repression of desires as a sign of evolutionary advance, sociobiologists and evolutionary psychologists view humans as driven by their prehistorical desires toward the maximization of reproductive fitness. Accordingly, Darwin's and evolutionary psychologists' narrative trajectories differ in their logics of change. For Darwin, the evolution of sexual morals is a key evolutionary event and the ultimate evidence of the inherent (although never secured) progress of evolution. Sociobiologists and evolutionary psychologists, by contrast, emphasize desire rather than its moderation, with the result that progress becomes equated with the reproductive sex act itself.

There is also considerable continuity between Darwin's, sociobiologists' and evolutionary psychologists' narratives of evolutionary change. Most importantly, they all posit reproduction as the organizing logic of evolution. This logic is fleshed out in extraordinary detail and with remarkable insistence in socio-biological and evolutionary psychological accounts of evolution. Yet the reproductive imperative also underlies more moderate and ecologically attuned versions of the evolutionary narrative such as Epic of Evolution texts, which appeared in the margins of, or in opposition to, sociobiology and evolutionary psychology. At the same time, the role ascribed to reproductive desire as the structural engine of evolutionary narratives varies from that of an unpredictable productive force in Darwin's texts to that of a fixed principle of organization in texts by such evolutionary psychologists as David Buss or Matt Ridley. As the cultural reception of different versions of evolutionary narratives suggests, slight shifts of emphasis may radically alter the sexual and cultural politics that emerge.

Furthermore, the relationship between sociobiology and evolutionary psychology is characterized by evolutionary psychologists' reworking of the narrative potential inherent in sociobiology. While both sets of narratives extend and often fetishize the reproductive imperative, evolutionary psycho-logical accounts tend to deploy less violent and more optimistic imagery than sociobiological narratives. The most significant difference between the two sets of narratives may be the role given to the gene and molecular imagery. Although introduced by Richard Dawkins in *The Selfish Gene* already in the 1970s, the gene as the evolutionary narrative subject and the true embodiment of agency became thoroughly integrated in popular evolutionary narratives with the introduction of the Human Genome Project in the 1990s. Thus, while Wilson's *On Human Nature* evokes the gene only occasionally and in a rhetorical manner, a text like Bryan Sykes' *Adam's Curse* premises its narrative dynamic on molecular agency. Crucially, the molecular evolutionary narrative has brought a sense of unity to evolutionary narratives by bridging the gap between the immensely slow species-level change and the destinies of individual organisms.

Structural continuities also underlie the different versions of evolutionary narrative that coexist in the same historical and cultural context. Discursive borders between different fields of cultural meaning-making always "leak" with the result that not only imagery or metaphor but also narrative elements travel from one field to another. For example, the molecular evolutionary narrative is regularly appropriated outside evolutionary psychology, as we saw in Natalie Angier's rewriting of the molecular microcosm as a utopian feminist narrative universe, and in Simon Mawer's framing of the molecular narrative as a site of an ethical dilemma. Throughout these intertextual travels, the molecular evolutionary narrative maintains its sense of interpretative primacy, thereby naturalizing gendered cultural imageries.

Formal affinities

While contemporary evolutionary narratives seek to invoke Darwin's narrative legacy, they also appropriate nonscientific cultural narratives. For example, both Epic of Evolution and philosophical naturalist texts develop the idea of a foundational narrative through religious discourse and narrative elements. Many accounts also rewrite evolution as an implicitly national narrative. This narrative framing equates scientific advance with national success, as in some philosophical naturalist texts and Darnton's novel, or depicts evolution as a narrative of the rise and fall of a nation, as in Will Self's novel. While these are not the only ways in which evolution can be imagined as a religious or national narrative, they reflect Darwin's ambiguity about foundationality and his invocation of spiritual awe and British imperialism.

The evolutionary infidelity narrative, on the other hand, dramatizes the central premise of evolution—the reproductive narrative logic—by portraying promiscuity and infidelity as logical outcomes of the reproductive imperative. While infidelity is not a necessary result of the adaptationist logic, it does find support in the event-based narrative dynamic of evolution. Most importantly, however, evolutionary promiscuity appears to confirm longstanding cultural assumptions about men and women as trapped in an eternal sex war. Narrative structure, this suggests, sets the limits of how far evolutionary narratives can be extended. At the same time, cultural context determines what can be understood as a narrative event, climax, or crisis in the first place. The evolutionary infidelity narrative resonates so well with cultural sensibilities because it appropriates both popular discourses of gender and a familiar narrative dynamic.

This contested interface between narrative form and cultural context is perhaps most evident in the structural parallelism between evolutionary and romantic narratives. I have argued that the narrative dynamic of romance provides evolutionary narratives with a structural defense against the possibility of reproductive failure by suggesting narrative continuity. As an implicit promise of narrative futurity, romantic narrative helps buttress the idea of evolution as an unending narrative trajectory. This structural alignment of two narrative dynamics also underlies evolutionary narratives' evocation of religious discourse. In philosophical naturalist accounts, for example, the familiar dynamic of fall and redemption operates as a means of supporting the adaptationist narrative logic. Through this structural parallelism, evolutionary accounts import assumptions of foundationality and cultural authority while rejecting the religious claims of creationism.

These structural affinities suggest that evolutionary narratives cannot be reduced either to the scientific context from which they emerged or to the cultural context that they reflect and reinforce. These affinities also indicate that there is considerable leakage across discursive fields and narrative structures. To an extent, this is the precondition of narrative intelligibility, the feeling that a story sounds like it should, that it matches with our prior knowledge of narratives. Crucially, this discursive and narrative leakage is also what makes it possible for the narrative engine of one narrative—romantic love and the need for self-expression, for example—to stand for the narrative engine of another narrative—that of evolution. This structural familiarity may be one reason why evolutionary accounts have proved so appealing across the cultural spectrum. In this sense, evolutionary narratives in fact rely on their structural resemblance to other narratives. At the same time, the resonances produced through structural similarities may also partly explain why evolutionary narratives have raised so many cultural anxieties.

While transgressing generic borders, narratives are not immune to the conventions of the genres in which they appear. Although narrative conventions tend to stabilize meaning, generic border crossing opens the borrowed textual elements for often unintended and unforeseen interpretations. The previous chapters have demonstrated that literary appropriations of evolutionary narrative may produce considerable ambiguity as to the authority of evolution as a cultural narrative, as is the case with Jenny Davidson's and Jeffrey Eugenides' novels. However, Lodge's and Darnton's novels demonstrate that there is nothing inherent in works of fiction that would necessitate such a critique.

Narrative as knowledge

This book has argued that evolutionary narratives are fundamentally implicated in questions of epistemic authority. As technologies of knowledge, narratives tend to render certain things sensible or obvious. At the same time, evolutionary narratives' focus on such foundational issues as origins, the categories of being (species, varieties, genders), the conditions of change, and the logics of movement connects them to larger metaphysical debates. Evolutionary narratives are accounts of how the world, including humanity, appeared, and what the principles governing this emergence of life are. They are also accounts of scientific progress, in which evolutionary theory appears as the narrative culmination and thus the privileged site from where evolutionary trajectories are to be judged. These two narratives are mutually intertwined so that the progress of one narrative (natural evolution) suggests the progress of the other (evolutionary theory as an epistemic enterprise), with the result that questions of ontology and epistemology appear as fundamentally entangled. The reproductive logic of the evolutionary narrative supports this association, as the two narratives become equated, respectively, with the production of narrative futurity (the survival of old life forms, the emergence of new species) and the production of knowledge about the mechanisms driving that future.

This entanglement of narrative and epistemology becomes evident through three cases discussed in the previous chapters. The first is that of philosophical naturalist texts that portray evolution as having led to scientific advance, which the philosophical naturalist understanding of evolution itself represents. While the narrative of scientific advance emerges from the narrative of evolutionary change, this advance also operates as proof of the validity of the evolutionary narrative from which it originated. This produces a curious circular movement in which scientific advance is used as evidence for the accuracy of the evolutionary narrative, which in turn is used as evidence for the superiority of the narrative of scientific advance. As we saw in Chapter 2, this sense of epistemic irrefutability is reinforced through the tone of divine necessity produced through religious references. The result is a textual economy in which narrative structure (the narrative of scientific advance) naturalizes discursive choices (the images of origins and destinies), while discursive choices (religious imagery) support claims implicit in narrative structure (evolution as a foundational narrative).

The second case is the evocation of scientific advance through reference to cutting-edge technology. As we saw in Chapter 3, molecular evolutionary narratives in particular posit advanced scientific and medical technologies— especially visual technologies—as the very condition of knowledge. This textual strategy excludes nonscientists as knowing subjects, rendering them passive recipients of the insights provided by the scientific enterprise. At the same time, science emerges as the only site of knowledge and the scientist as the epistemically privileged subject who, unlike the nonscientist, is able to rise above the fuzzy and confused communication among the rest of the humanity. In contrast to men and women blinded by their prehistoric proclivities, the evolutionary biologist is able to see. This rhetorical strategy connects the narrative of evolution to the narrative of scientific advance through the valorized image of visual technology, thereby establishing the molecular visions it claims to produce as the site where the truth is located. Such narrative politics strengthen the idea of science in general and evolutionary biology in particular as the single source of insight into human existence and its historical roots.

The third case is the representation of nonreproductive sexualities in evolutionary narratives. As we saw in Chapter 5, the adaptationist reproductive logic that organizes evolutionary psychological narratives renders nonreproductive sexuality an apparent theoretical challenge. Instead of threatening the adaptationist narrative, however, nonreproductive sexualities provide the fundamental element of crisis through which adaptationist evolutionary narratives can establish themselves as epistemically privileged accounts of gender and sexuality. Like visual technologies, nonreproductive sexualities serve as a link between the narrative of evolutionary change and the narrative of scientific progress. By claiming to have explained such presumably deviant sexualities, the narrative of science proves that it is indeed a narrative of progress.

These three examples suggest that shifts in the relationship between narrative structure and cultural context always carry epistemic consequences. By evoking structural affinities and discursive affiliations available in the given historical context, evolutionary narratives may represent themselves as epistemically authoritative accounts of the world. There are, however, considerable differences between the kinds of elements that evolutionary narratives appropriate. Furthermore, different versions of the narrative—say, adaptationist thought versus Epic of Evolution—differ in their cultural visibility. This suggests that certain strategies, such as the gene-centered narrative logic of evolutionary psychology, may appear as more rhetorically appealing to contemporary cultural audiences. At the same time, there is no form of evolutionary narrative that may reside outside the workings of ideology. By the very fact of its organizing structure, an evolutionary narrative is always also an epistemic claim.

The location of resistance

This book has argued that an analysis of the changing destinies of the evolutionary narrative is able to locate and challenge problematic assumptions about gender, sexuality, and human nature in contemporary evolutionary discourse. The previous chapters have pointed to inherent contradictions and structural weaknesses in the seemingly infallible logic of evolutionary narration. However, my aim has not been to question the validity of evolutionary theory as a scientific practice but to suggest that evolutionary models—like all models—have their limitations in interpreting the vast and evasive field of human embodiment and experience.

The evolutionary narratives this book has traced carry elements that undermine their coherence as cultural narratives. Perhaps the most central structural weakness is the narrative reliance on reproductive success, which is never fully guaranteed. This narrative instability is masked by the representation of the reproductive imperative as a force driving genes, organisms, and narrative, as well as through the invocation of cultural narratives that suggest continuity. Yet the persistent presence of infertility imagery suggests that evolutionary narratives are never fully able to contain the possibility of reproductive failure. Similarly, the idea of evolution as a never-ending line of ascent is an effect of the retrospective nature of evolutionary knowledge. It is only as told from the survivor's perspective that evolutionary narratives can appear as linear and secured.

Furthermore, the adaptationist logic that lies at the heart of evolutionary narratives in general and sociobiological and evolutionary psychological narratives in particular understands change in terms of acts. This event-based logic reinterprets a wide field of experience—sensations, pleasures, desires, and identities—as reproductively motivated narrative events, or reduces them into mere side effects of reproductive ambitions. The way in which evolutionary narratives often evoke the image of an eon-spanning reproductive continuum tends to render invisible these reductions and exclusions. Yet the inability of the adaptationist logic to account for phenomena that are irreducible to distinct, isolated acts undermines the explanatory scope of evolutionary narratives. While such haunting is indicative of the multiplicity of sexual experience, it is also symptomatic of the excess and evasiveness of cultural phenomena at large. Culture, that is, is simply too vast, too complex, too independent to be captured within any single narrative logic, evolutionary or not. By refusing to read the multitude of experience through the adaptationist logic of narrative events, then, we may be able to make space for rewriting and revision.

Finally, the structural similarities between evolutionary narratives and other cultural narratives suggest that narrative structure cannot be pinned unambiguously to any single ideology. As the examples of national narrative, religious narrative, and romantic narrative show, the narrative structure of Darwinian evolution resonates with more than one discursive framework or ideological position. This implies that the relationship between narrative structure and cultural context is always ambiguous, never fixed, constantly suggesting more than either part of the equation—narrative or culture—can contain. Evolutionary explanation produces narratives that are not nearly as all-encompassing as they often claim to be. While proponents of evolutionary psychology emphasize stable gender identities, underlying sexual natures and a neat linear historical trajectory, this politics of stability and movement is underwritten by considerable narrative instability. It is only through constant struggle to contain narrative tension and underlying contradictions that evolutionary narratives produce the image of secured continuity and epistemic authority they are often associated with in contemporary culture. It is, then, in this contested zone between narrative structure and cultural context that feminist and queer resistance may arise.

Bibliography

Abbott, H. P. (2001) "Humanists, scientists, and the cultural surplus," *SubStance*, 30(1–2): 203–19.

——(2003) "Unnarratable knowledge: the difficulty of understanding evolution by natural selection," in D. Herman (ed.) *Narrative Theory and the Cognitive Sciences*, Stanford: CSLI Publications.

Adriaens, P. R. and De Block, A. (2006) "The evolution of a social construction: the case of male homosexuality," *Perspectives in Biology and Medicine*, 49(4): 570–85.

Albury, W. R. (1980) "Politics and rhetoric in the sociobiology debate," *Social Studies of Science*, 10(4): 519–36.

Allen, D. W. (1995) "Homosexuality and narrative," *Modern Fiction Studies*, 41(3–4): 609–34.

Allen, E., Beckwith, B., Beckwith, J., Chorover, S., Culver, D., Duncan, M., Gould, S., Hubbard, R., Inouye, H., Leeds, A., Lewontin, R., Madansky, C., Miller, L., Pyeritz, R., Rosenthal, M., and Schreier, H. (1975) "Against 'sociobiology'," *The New York Review of Books* (November 13). Online. Available HTTP: www.nybooks. com/articles/archives/1975/nov/13/against-sociobiology/ (accessed June 18, 2012).

Anderson, A. (2004) *Darwin's Wink: a novel of nature and love*, New York: Thomas Dunne Books/St. Martin's Press.

Angier, N. ([1999] 2000) *Woman: an intimate geography*, New York: Anchor Books.

Antolin, M. F. and Herbers J. M. (2001) "Evolution's struggle for existence in America's public schools," *Evolution*, 55(12): 2379–88.

Ayala, F. J. (2007) *Darwin's Gift to Science and Religion*, Washington, DC: Joseph Henry Press.

Bailey, J. M. and Pillard, R. (1991) "A genetic study of male sexual orientation," *Archives of General Psychiatry*, 48(12): 1089–96.

Baird, R. J. (2000) "Late secularism," *Social Text*, 18(3): 123–36.

Baker, R. ([1996] 2006) *Sperm Wars: infidelity, sexual conflict, and other bedroom battles*, New York: Thunder's Mouth Press.

Barash, D. P. (1979) *The Whisperings Within: evolution and the origin of human nature*, New York: Harper & Row.

Barkow, J. H., Cosmides, L., and Tooby, J. (1992) *The Adapted Mind: evolutionary psychology and the generation of culture*, Oxford: Oxford University Press.

Beer, G. ([1983] 2000) *Darwin's Plots: evolutionary narrative in Darwin, George Eliot and nineteenth-century fiction*, second edition, Cambridge: Cambridge University Press.

——(1999) *Open Fields: science in cultural encounter*, Oxford: Oxford University Press.

Berman, L. A. (2003) *The Puzzle: exploring the evolutionary puzzle of male homosexuality*, Wilmette, IL: Godot Press.

Bethell, T. (2001) "Against sociobiology," *First Things* (January). Online. Available HTTP: www.firstthings.com/issue/2001/01/january (accessed June 8, 2012).

Birkhead, T. (2000) *Promiscuity: an evolutionary history of sperm competition*, Cambridge, MA: Harvard University Press.

Bleier, R. (1984) *Science and Gender: a critique of biology and its theories on women*, New York: Pergamon Press.

Bowler, P. J. (2007) *Monkey Trials and Gorilla Sermons: evolution and Christianity from Darwin to intelligent design*, Cambridge, MA: Harvard University Press.

Burns, A. (2002) "Women in love and men at work: the evolving heterosexual couple?" *Psychology, Evolution & Gender*, 4(2): 149–72.

Buss, D. M. ([1994] 2003) *The Evolution of Desire: strategies of human mating*, revised edition, New York: Basic Books.

Butler, J. (2004) *Undoing Gender*, New York: Routledge.

Cannadine, D. (2005) "'Big tent' historiography: transatlantic obstacles and opportunities in writing the history of empire," *Common Knowledge*, 11(3): 375–92.

Castle, T. (1993) *The Apparitional Lesbian: female homosexuality and modern culture*, New York: Columbia University Press.

Ceccarelli, L. (2004) "Neither confusing cacophony nor culinary complements: a case study of mixed metaphors for genomic science," *Written Communication*, 21(1): 92–105.

Chaisson, E. (2005) *Epic of Evolution: seven ages of the cosmos*, New York: Columbia University Press.

Clayton, J. (1989) "Narrative and theories of desire," *Critical Inquiry*, 16(1): 33–53.

——(2002) "Genome time," in K. Newman, J. Clayton, and M. Hirsch (eds) *Time and the Literary*, New York: Routledge.

Collado-Rodríguez, F. (2006) "Of self and country: U.S. politics, cultural hybridity, and ambivalent identity in Jeffrey Eugenides's *Middlesex*," *The International Fiction Review*, 33(1–2): 71–83.

Conrad, P. and Markens, S. (2001) "Constructing the 'gay gene' in the news: optimism and skepticism in the US and British press," *Health*, 5(3): 373–400.

Cox, S. M. and McKellin, W. (1999) "'There's this thing in our family': predictive testing and the construction of risk for Huntington disease," *Sociology of Health & Illness*, 21: 622–46.

Curtis, R. (1994) "Narrative form and normative force: Baconian story-telling in popular science," *Social Studies of Science*, 24(3): 419–61.

Dale, R. (dir.) (2003) *Walking with Cavemen*, BBC, Discovery Channel and Pro-Sieben.

Darnton, J. ([1996] 1997) *Neanderthal*, New York: St. Martin's Press.

Darwin, C. ([1859] 1985) *The Origin of Species by Means of Natural Selection or the Preservation of Favoured Races in the Struggle for Life*, ed. J. W. Burrow, London: Penguin Books.

——([1868] 2010) *The Variation of Animals and Plants under Domestication*, vols 1–2, Cambridge: Cambridge University Press.

——([1879] 2004) *The Descent of Man, and Selection in Relation to Sex*, second edition, ed. J. Moore and A. Desmond, London: Penguin Books.

Davidson, J. (2003) *Heredity*, London: Serpent's Tail.

Dawkins, R. ([1976] 1999) *The Selfish Gene*, Oxford: Oxford University Press.

——([1982] 1999) *The Extended Phenotype: the long reach of the gene*, revised edition, Oxford: Oxford University Press.

——([1995] 1996) *River Out of Eden: a Darwinian view of life*, London: Phoenix.

——(2008) *The God Delusion*, Boston: Mariner Books/Houghton Mifflin.

de Lauretis, T. (1984) *Alice Doesn't: feminism, semiotics, cinema*, Bloomington: Indiana University Press.

Dennett, D. C. (1996) *Darwin's Dangerous Idea: evolution and the meanings of life*, London: Penguin Books.

——(2007) *Breaking the Spell: religion as a natural phenomenon*, London: Penguin Books.

Dewsbury, D. A. (2005) "The Darwin-Bateman paradigm in historical context," *Integrative and Comparative Biology*, 45(5): 831–7.

Dreger, A. D. (1998) *Hermaphrodites and the Medical Invention of Sex*, Cambridge, MA: Harvard University Press.

Driscoll, E. V. (2009) "Bisexual species," *Scientific American Mind*, 20(3): 20–5.

DuPlessis, R. B. (1985) *Writing beyond the Ending: narrative strategies of twentieth-century women writers*, Bloomington: Indiana University Press.

Dusek, V. (1999) "Sociobiology sanitized: evolutionary psychology and gene selectionism," *Science as Culture*, 8(2): 129–69.

The Economist (1998) "Sex, Death and Football," editorial, *The Economist* (June 13): 18.

Edelman, L. (2004) *No Future: queer theory and the death drive*, Durham, NC: Duke University Press.

Eldredge, N. (1985) *Time Frames: the rethinking of Darwinian evolution and the theory of punctuated equilibria*, New York: Simon and Schuster.

——(2004) *Why We Do It: rethinking sex and the selfish gene*, New York: W. W. Norton.

Eldredge, N. and Gould, S. J. (1972) "Punctuated equilibria: an alternative to phyletic gradualism," in T. J. M. Schopf (ed.) *Models in Paleobiology*, San Francisco: Freeman Cooper.

Epstein, R. (2009) "Do gays have a choice?" *Scientific American Mind*, 20(3): 62–9.

Eugenides, J. (2002) *Middlesex*, New York: Picador.

Fahnestock, J. (2004) "Preserving the figure: consistency in the presentation of scientific arguments," *Written Communication*, 21(1): 6–31.

Fausto-Sterling, A. (2000) *Sexing the Body: gender politics and the construction of sexuality*, New York: Basic Books.

Fausto-Sterling, A., Gowaty, P. A., and Zuk, M. (1997) "Evolutionary psychology and Darwinian feminism," *Feminist Studies*, 23(2): 403–17.

Farwell, M. R. (1996) *Heterosexual Plots and Lesbian Narratives*, New York: New York University Press.

Felski, R. (2003) *Literature after Feminism*, Chicago and London: The University of Chicago Press.

Fisher, H. (1994) *Anatomy of Love: a natural history of mating, marriage, and why we stray*, New York: Fawcett/Ballantine Books.

Fogle, T. (1995) "Information metaphors and the human genome project," *Perspectives in Biology and Medicine*, 38(4): 535–47.

Franklin, S. (1995) "Romancing the helix: nature and scientific discovery," in L. Pearce and J. Stacey (eds) *Romance Revisited*, New York: New York University Press.

Friedman, S. S. (1998) *Mappings: feminism and the cultural geographies of encounter*, Princeton, NJ: Princeton University Press.

Genet, R. M. (1998) "The epic of evolution: a course developmental project," *Zygon*, 33(4): 635–44.

Gieryn, T. F. (1999) *Cultural Boundaries of Science: credibility on the line*, Chicago: The University of Chicago Press.

Gilbert, M. (2006) *The Disposable Male: sex, love, and money: your world through Darwin's eyes*, Atlanta, GA: The Hunter Press.

Gill, R. and Herdieckerhoff, E. (2006) "Rewriting the romance: new femininities in chick lit?" *Feminist Media Studies*, 6(4): 487–504.

Goodenough, U. (1998) *The Sacred Depths of Nature*, Oxford: Oxford University Press.

Gould, S. J. ([1981] 1996) *The Mismeasure of Man*, New York: W. W. Norton.

——([1989] 1990) *Wonderful Life: the Burgess Shale and the nature of history*, London: Hutchinson Radius.

——(1999) *Rocks of Ages: science and religion in the fullness of life*, New York: Ballantine Books.

Gould, S. J. and Lewontin, R. (1979) "The spandrels of San Marco and the panglossian paradigm: a critique of the adaptationist programme," *Proceedings of the Royal Society of London*, (Series B) 205(1161): 581–98.

Gray, J. (1992) *Men Are from Mars, Women Are from Venus*, London: Thorsons.

Gray, J. P. and Wolfe, L. D. (1982) "Sociobiology and creationism: two ethnosociologies of American culture," *American Anthropologist*, 84(3): 580–94.

Gregory, J. and Miller, S. (1998) *Science in Public: communication, culture, and credibility*, Cambridge, MA: Basic Books.

Grosz, E. (2004) *The Nick of Time: politics, evolution, and the untimely*, Durham, NC: Duke University Press.

——(2005) *Time Travels: feminism, nature, power*, Durham, NC: Duke University Press.

——(2007) "Feminism, art, Deleuze, and Darwin: an interview with Elizabeth Grosz," conducted by Katve-Kaisa Kontturi and Milla Tiainen, *NORA: Nordic Journal of Women's Studies*, 15(4): 246–56.

Gudding, G. (1996) "The phenotype/genotype distinction and the disappearance of the body," *Journal of the History of Ideas*, 57(3): 525–45.

Hall, S. S. (2008) "Last of the Neanderthals," *National Geographic Magazine* (October). Online. Available HTTP: http://ngm.nationalgeographic.com/2008/10/neanderthals/hall-text (accessed August 20, 2012).

Hamer, D. and Copeland, P. (1994) *The Science of Desire: the search for the gay gene and the biology of behavior*, New York: Simon & Schuster.

Hamer, D. H., Hu, S., Magnuson, V. L., Hu, N., and Pattatucci, A. M. (1993) "A linkage between DNA markers on the X chromosome and male sexual orientation," *Science*, 261(5119): 321–7.

Haraway, D. J. (1989) *Primate Visions: gender, race, and nature in the world of modern science*, New York: Routledge.

——(1991) *Simians, Cyborgs, and Women: the reinvention of nature*, London: Free Association Books.

——(1997) *Modest_Witness@Second_Millennium.FemaleMan©_Meets_OncoMouse^{TM}*, New York: Routledge.

Hausman, B. L. (2000) "Do boys have to be boys?: gender, narrativity, and the John/Joan case," *NWSA Journal*, 12(3): 114–38.

Hayles, N. K. (2001) "Desiring agency: limiting metaphors and enabling constraints in Dawkins and Deleuze/Guattari," *SubStance*, 30(1–2): 144–59.

Hedgecoe, A. M. (1999) "Transforming genes: metaphors of information and language in modern genetics," *Science as Culture*, 8(2): 209–29.

Hemmings, C. (2011) *Why Stories Matter: the political grammar of feminist theory*, Durham, NC: Duke University Press.

Hilgartner, S. (1990) "The dominant view of popularization: conceptual problems, political uses," *Social Studies of Science*, 20(3): 519–39.

Homans, M. (1994) "Feminist fictions and feminist theories of narrative," *Narrative*, 2(1): 3–16.

Hrdy, S. B. (1981) *The Woman that Never Evolved*, Cambridge, MA: Harvard University Press.

——(1999) *Mother Nature: maternal instincts and how they shape the human species*, New York: Ballantine Books.

Hsu, S. (2011) "Ethnicity and the biopolitics of intersex in Jeffrey Eugenides's *Middlesex*," *MELUS*, 36(3): 87–110.

Humes, E. (2008) *Monkey Girl: evolution, education, religion, and the battle for America's soul*, New York: Harper Perennial.

Irons, P. (2007) *God on Trial: dispatches from America's religious battlefields*, New York: Viking.

Jenkins, K. E. (2007) "Genetics and faith: religious enchantment through creative engagement with molecular biology," *Social Forces*, 85(4): 1693–712.

Joyce, K. A. (2006) "From numbers to pictures: the development of magnetic resonance imaging and the visual turn in medicine," *Science as Culture*, 15(1): 1–22.

Judson, O. ([2002] 2003) *Dr. Tatiana's Sex Advice to All Creation: the definitive guide to the evolutionary biology of sex*, London: Vintage.

Karkazis, K. (2008) *Fixing Sex: intersex, medical authority, and lived experience*, Durham, NC: Duke University Press.

Keller, E. F. (1995) *Refiguring Life: metaphors of twentieth-century biology*, New York: Columbia University Press.

——(2000) *The Century of the Gene*, Cambridge, MA: Harvard University Press.

Lancaster, R. N. (2003) *The Trouble with Nature: sex in science and popular culture*, Berkeley: University of California Press.

——(2006) "Sex, science, and pseudoscience in the public sphere," *Identities: Global Studies in Culture and Power*, 13(1): 101–38.

Landau, M. (1991) *Narratives of Human Evolution*, New Haven, CT: Yale University Press.

Lanser, S. S. (1996) "Queering narratology," in K. Mezei (ed.) *Ambiguous Discourse: feminist narratology and British women writers*, Chapel Hill: The University of North Carolina Press.

——(2009) "Novel (lesbian) subjects: the sexual history of form," *Novel: A Forum on Fiction*, 42(3): 497–503.

Larson, E. J. ([2004] 2006) *Evolution: the remarkable history of a scientific theory*, New York: Modern Library.

Lemonick, M. D. (2003) "A twist of fate," *Time* (February 17). Online. Available HTTP: www.time.com/time/magazine/article/0,9171,1004241,00.html (accessed September 19, 2012).

——(2006) "What makes us different?" *Time* (October 1). Online. Available HTTP: www.time.com/time/magazine/article/0,9171,1541283,00.html (accessed August 20, 2012).

Lennox, J. G. (1993) "Darwin *was* a teleologist," *Biology and Philosophy*, 8(4): 409–21.

Lessl, T. M. (2002) "Gnostic scientism and the prohibition of questions," *Rhetoric and Public Affairs*, 5(1): 133–57.

LeVay, S. (1991) "A difference in hypothalamic structure between homosexual and heterosexual men," *Science*, 253(5023): 1034–7.

Levine, G. (2011) *Darwin the Writer*, Oxford: Oxford University Press.

Lewontin, R. C., Rose, S., and Kamin, L. J. (1984) *Not in Our Genes: biology, ideology, and human nature*, New York: Pantheon Books.

Lloyd, E. A. (2005) *The Case of the Female Orgasm: bias in the science of evolution*, Cambridge, MA: Harvard University Press.

Lodge, D. ([2001] 2002) *Thinks …* , London: Penguin Books.

Majdik, Z. P. (2009) "Judging direct-to-consumer genetics: negotiating expertise and agency in public biotechnological practice," *Rhetoric & Public Affairs*, 12(4): 571–605.

Marchessault, J. (2000) "David Suzuki's *The Secret of Life*: informatics and the popular discourse of the life code," in J. Marchessault and K. Sawchuk (eds) *Wild Science: reading feminism, medicine and the media*, London: Routledge.

Margulis, L. (1998) *Symbiotic Planet: a new look at evolution*, New York: Basic Books.

Margulis, L. and Sagan, D. ([1986] 1997) *Microcosmos: four billion years of microbial evolution*, revised edition, Berkeley and Los Angeles: University of California Press.

Martin, E. (1991) "The egg and the sperm: how science has constructed a romance based on stereotypical male-female roles," *Signs: Journal of Women in Culture and Society*, 16(3): 485–501.

Mawer, S. (1998) *Mendel's Dwarf*, New York: Harmony Books.

McCaughey, M. (2008) *The Caveman Mystique: pop-Darwinism and the debates over sex, violence, and science*, New York and London: Routledge.

McGrath, A. (2007) *Dawkins' God: genes, memes, and the meaning of life*, Malden, MA: Blackwell Publishing.

McKinnon, S. (2006) *Neo-liberal Genetics: the myths and moral tales of evolutionary psychology*, Chicago: Prickly Paradigm Press.

Mellor, F. (2003) "Between fact and fiction: demarcating science from non-science in popular physics books," *Social Studies of Science*, 33(4): 509–38.

Midgley, M. ([1979] 1995) *Beast and Man: the roots of human nature*, revised edition, London and New York: Routledge.

Milburn, C. N. (2003) "Monsters in Eden: Darwin and Derrida," *MLN*, 118(3): 603–21.

Miller, G. ([2000] 2001) *The Mating Mind: how sexual choice shaped the evolution of human nature*, New York: Anchor Books.

Miller, K. R. (2007) *Finding Darwin's God: a scientist's search for common ground between God and evolution*, New York: Harper Perennial.

Modleski, T. (1984) *Loving with a Vengeance: mass-produced fantasies for women*, New York: Methuen.

Money, J. (1986) *Venuses Penuses: sexology, sexosophy, and exigency theory*, Amherst, NY: Prometheus Books.

Montgomery, S. L. (1996) *The Scientific Voice*, New York: The Guilford Press.

Moore, J. and Desmond, A. (2004) "Introduction," in C. Darwin ([1879] 2004) *The Descent of Man, and Selection in Relation to Sex*, second edition, ed. J. Moore and A. Desmond, London: Penguin Books.

Morgan, E. ([1972] 2001) *The Descent of Woman: the classic study of evolution*, London: Souvenir Press.

Morland, I. (2001) "Is intersexuality real?" *Textual Practice*, 15(3): 527–47.

Morris, D. (1967) *The Naked Ape: a zoologist's study of the human animal*, New York: McGraw-Hill.

Muñoz-Rubio, J. (2003) "Charles Darwin: continuity, teleology and ideology," *Science as Culture*, 12(3): 303–39.

Myers, G. (1990) *Writing Biology: texts in the social construction of scientific knowledge*, Madison: The University of Wisconsin Press.

——(2003) "Discourse studies of scientific popularization: questioning the boundaries," *Discourse Studies*, 5(2): 265–79.

Nee, S. (2005) "The great chain of being," *Nature*, 435(7041): 429.

Nelkin, D. and Lindee, M. S. ([1996] 2004) *The DNA Mystique: the gene as a cultural icon*, second edition, Ann Arbor: University of Michigan Press.

O'Hara, R. J. (1988) "Homage to Clio, or, toward an historical philosophy for evolutionary biology," *Systematic Zoology*, 37(2): 142–55.

——(1992) "Telling the tree: narrative representation and the study of evolutionary history," *Biology and Philosophy*, 7(2): 135–60.

Ospovat, D. (1980) "God and natural selection: the Darwinian idea of design," *Journal of the History of Biology*, 13(2): 169–94.

Page, R. (2007) "Gender," in D. Herman (ed.) *The Cambridge Companion to Narrative*, Cambridge: Cambridge University Press.

Parisi, L. (2010) "Event and evolution," *The Southern Journal of Philosophy*, 48: 147–64.

Paul, D. (2004) "Spreading chaos: the role of popularization in the diffusion of scientific ideas," *Written Communication*, 21(1): 32–68.

Peters, N. J. (2006) *Conundrum: the evolution of homosexuality*, Bloomington, IN: AuthorHouse.

Pfeiffer, J. (1975) "Sociobiology," *New York Times* (July 27): BR4.

Pinker, S. (1997) *How the Mind Works*, New York and London: W. W. Norton.

——(2002) *The Blank Slate: the modern denial of human nature*, New York: Viking.

Potts, A. (2002) *The Science/Fiction of Sex: feminist deconstruction and the vocabularies of heterosex*, London: Routledge.

Potts, M. and Short, R. (1999) *Ever Since Adam and Eve: the evolution of human sexuality*, Cambridge: Cambridge University Press.

Radway, J. A. (1991) *Reading the Romance: women, patriarchy, and popular literature*, Chapel Hill: The University of North Carolina Press.

Reed, E. (1978) *Sexism and Science*, New York: Pathfinder Press.

Rees, A. (2007) "Reflections on the field: primatology, popular science and the politics of personhood," *Social Studies of Science*, 37(6): 881–907.

Regis, P. (2003) *A Natural History of the Romance Novel*, Philadelphia: University of Pennsylvania Press.

Reis, E. (2009) *Bodies in Doubt: an American history of intersex*, Baltimore, MD: Johns Hopkins University Press.

Rensberger, B. (1975) "Sociobiology: updating Darwin on behavior," *New York Times* (May 28): 1, 52.

Ridley, M. ([1993] 2003) *The Red Queen: sex and the evolution of human nature*, New York: Harper Perennial.

——(2009) "Modern Darwins," *National Geographic*, 215(2) (February): 56–73.

Roof, J. (1996) *Come as You Are: sexuality and narrative*, New York: Columbia University Press.

——(2007) *The Poetics of DNA*, Minneapolis: University of Minnesota Press.

Rose, H. and Rose, S. (2001) *Alas, Poor Darwin: arguments against evolutionary psychology*, London: Vintage.

Rose, N. (2007) *The Politics of Life Itself: biomedicine, power, and subjectivity in the twenty-first century*, Princeton, NJ: Princeton University Press.

Roughgarden, J. (2004) *Evolution's Rainbow: diversity, gender, and sexuality in nature and people*, Berkeley: University of California Press.

Rue, L. (2000) *Everybody's Story: wising up to the epic of evolution*, Albany: State University of New York Press.

Ruse, M. (2006) *The Evolution-Creation Struggle*, Cambridge, MA: Harvard University Press.

Schell, H. (2007) "The big bad wolf: masculinity and genetics in popular culture," *Literature and Medicine*, 26(1): 109–25.

Schiebinger, L. ([1993] 2004) *Nature's Body: gender in the making of modern science*, revised edition, New Brunswick, NJ: Rutgers University Press.

Scott, E. C. (2005) *Evolution vs. Creationism: an introduction*, Berkeley: University of California Press.

Segerstråle, U. (2000) *Defenders of the Truth: the sociobiology debate*, Oxford: Oxford University Press.

Self, W. (1997) *Great Apes*, New York: Grove Press.

Selinger, E. M. (2007) "Rereading the romance," *Contemporary Literature*, 48(2): 307–24.

Shea, E. (2001) "The gene as a rhetorical figure: 'nothing but a very applicable little word'," *Science as Culture*, 10(4): 505–29.

Shostak, D. (2008) "'Theory uncompromised by practicality': hybridity in Jeffrey Eugenides' *Middlesex*," *Contemporary Literature*, 49(3): 383–412.

Sifuentes, Z. (2006) "Strange anatomy, strange sexuality: the queer body in Jeffrey Eugenides' *Middlesex*," in R. Fantina (ed.) *Straight Writ Queer: non-normative expressions of heterosexuality in literature*, Jefferson, NC: McFarland.

Slack, G. (2007) *The Battle over the Meaning of Everything: evolution, intelligent design, and a school board in Dover, PA*, San Francisco: Jossey Bass.

Spencer, H. (1864) *The Principles of Biology*, vol. 1, London and Edinburgh: Williams and Norgate.

Sperling, S. (1991) "Baboons with briefcases: feminism, functionalism, and sociobiology in the evolution of primate gender," *Signs: Journal of Women in Culture and Society*, 17(1): 1–27.

Steintrager, J. A. (1999) "'Are you there yet?': libertinage and the semantics of the orgasm," *differences: A Journal of Feminist Cultural Studies*, 11(2): 22–52.

Stenger, V. J. (2008) *God: the failed hypothesis – how science shows that god does not exist*, New York: Prometheus Books.

Swimme, B. and Berry, T. (1992) *The Universe Story: from the primordial flaring forth to the ecozoic era – a celebration of the unfolding of the cosmos*, New York: HarperOne.

Sykes, B. ([2003] 2004) *Adam's Curse: A Future without Men*, New York: W. W. Norton.

Taub, S., Morin, K., Spillman, M. A., Sade, R. M., and Riddick, F. A. (2004) "Managing familial risk in genetic testing," *Genetic Testing*, 8: 356–9.

Tennyson, A. ([1908] 2003) *In Memoriam*, second edition, ed. Erik Gray, New York: W. W. Norton.

Terry, J. (2000) "'Unnatural acts' in nature: the scientific fascination with queer animals," *GLQ: A Journal of Lesbian and Gay Studies*, 6(2): 151–93.

Thacker, E. (2003) "Data made flesh: biotechnology and the discourse of the posthuman," *Cultural Critique*, 53: 72–97.

Than, K. (2010) "Neanderthals, humans interbred – first solid DNA evidence," *National Geographic* (May 6). Online. Available HTTP: http://news.nationalgeographic.com/news/2010/05/100506-science-neanderthals-humans-mated-interbred-dna-gene/ (accessed August 20, 2012).

Torgersen, H. (2009) "Fuzzy genes: epistemic tensions in genomics," *Science as Culture*, 18(1): 65–87.

Tourney, C. P. (1991) "Modern creationism and scientific authority," *Social Studies of Science*, 21(4): 681–99.

Trivers, R. L. (1972) "Parental investment and sexual selection," in B. G. Campbell (ed.) *Sexual Selection and the Descent of Man*, Chicago: Aldine.

Tuana, N. (2004) "Coming to understand: orgasm and the epistemology of ignorance," *Hypatia*, 19(1): 194–232.

Turner, S. S. (2007) "Open-ended stories: extinction narratives in genome time," *Literature and Medicine*, 26(1): 55–82.

Turney, J. (1999) "The word and the world: engaging with science in print," in E. Scanlon, E. Whitelegg, and S. Yates (eds) *Communicating Science: contexts and channels*, New York: Routledge/The Open University.

——(2001) "Telling the facts of life: cosmology and the epic of evolution," *Science as Culture*, 10(2): 225–47.

van Dijck, J. (1998) *Imagenation: popular images of genetics*, Houndmills: Macmillan.

——(2000) "The language and literature of life: popular metaphors in genome research," in J. Marchessault and K. Sawchuk (eds) *Wild Science: reading feminism, medicine and the media*, London: Routledge.

Varghese, S. A. and Abraham, S. A. (2004) "Book-length scholarly essays as a hybrid genre in science," *Written Communication*, 21(2): 201–31.

Vassey, P. L. (1998) "Intimate sexual relations in prehistory: lessons from the Japanese macaques," *World Archaeology*, 29(3): 407–25.

Wade, N. (1997) "Male chromosome is not a genetic wasteland, after all," *The New York Times* (October 28). Online. Available HTTP: www.nytimes.com/1997/10/28/science/male-hromosome-is-not-a-genetic-wasteland-after-all.html (accessed September 19, 2012).

——(2003) "Y chromosome depends on itself to survive," *The New York Times* (June 19). Online. Available HTTP: www.nytimes.com/2003/06/19/us/y-chromosome-depends-on-itself-to-survive.html (accessed September 19, 2012).

——(2007) "Pas de deux of sexuality is written in the genes," *The New York Times* (April 10). Online. Available HTTP: www.nytimes.com/2007/04/10/health/10gene.html (accessed October 22, 2008).

Waddington, C. H. (1975) "Mindless societies," *The New York Review of Books* (August 7). Online. Available HTTP: www.nybooks.com/articles/archives/1975/aug/07/mindless-societies/ (accessed June 18, 2012).

Wald, P. (2000) "Future perfect: grammar, genes, and geography," *New Literary History*, 31(4): 681–708.

——(2005) "What's in a cell?: John Moore's spleen and the language of bioslavery," *New Literary History*, 36(2): 205–25.

——(2008) *Contagious: cultures, carriers, and the outbreak narrative*, Durham, NC: Duke University Press.

Waldby, C. (1997) "Revenants: the visible human project and the digital uncanny," *Body & Society*, 3(1): 1–16.

Wallace, B. (2010) *Getting Darwin Wrong: why evolutionary psychology won't work*, Exeter and Charlottesville, VA: Imprint Academic.

Wallis, C. (2005) "The evolution wars," *Time* (August 7). Online. Available HTTP: www.time.com/time/magazine/article/0,9171,1090909,00.html (accessed February 25, 2009).

Watson, J. D. and Crick, F. H. C. (1953) "A structure for deoxyribose nucleic acid," *Nature*, 171(4356) (April 25): 737–8.

Waugh, P. (2005) "Just-so stories: science, narrative, and postmodern intertextualities," *Symbolism: An International Annual of Critical Aesthetics*, 5: 223–63.

Weinstein, J. (2010) "A requiem to sexual difference: a response to Luciana Parisi's 'Event and evolution'," *The Southern Journal of Philosophy*, 48: 165–87.

Weiss, M. (2005) "Court battle over teaching of evolution," *San Francisco Chronicle* (November 6). Online. Available HTTP: http://sfgate.com/cgi-bin/article.cgi?f=/c/a/2005/11/06/DESIGN.TMP (accessed February 25, 2009).

Wilcox, S. A. (2003) "Cultural context and the conventions of science journalism: drama and contradiction in media coverage of biological ideas about sexuality," *Critical Studies in Media Communication*, 20(3): 225–47.

Williams, P. A. (2001) *Doing without Adam and Eve: sociobiology and original sin*, Minneapolis, MN: Fortress Press.

Wilson, E. O. ([1975] 1982) *Sociobiology: the new synthesis*, Cambridge, MA and London: The Belknap Press of Harvard University Press.

——(1978) *On Human Nature*, Cambridge, MA: Harvard University Press.

——([1998] 2001) *Consilience: the unity of knowledge*, London: Abacus.

Winnett, S. (1990) "Coming unstrung: women, men, narrative, and principles of pleasure," *PMLA*, 105(3): 505–18.

Winston, R. P. and Marshall, T. (2002) "The shadows of history: the 'condition of England' in *Nice Work*," *Critique*, 44(1): 3–22.

Wong, K. (2009) "Twilight of the Neandertals," *Scientific American* (August): 34–9.

Wright, R. ([1994] 2004) *The Moral Animal: why we are the way we are*, London: Abacus.

Young, R. M. (1985) *Darwin's Metaphor: nature's place in Victorian culture*, Cambridge: Cambridge University Press.

Zimmer, C. (2008) "Romance is an illusion," *Time* (January 28). Online. Available HTTP: www.time.com/time/magazine/article/0,9171,1704665,00.html (accessed December 10, 2008).

Žižek, S. (2002) "Cultural studies versus the 'third culture'," *The South Atlantic Quarterly*, 101(1): 19–32.

Zuk, M. (2002) *Sexual Selections: what we can and can't learn about sex from animals*, Berkeley: University of California Press.

Index

Abbott, H. Porter 23, 69, 81, 82, 105, 127, 128, 129, 146

Abraham, Sunita Anne 31

Adam's Curse: A Future without Men (Sykes, B.) 15, 68, 116, 130, 156; gendered politics of genetic discourse 77, 78–80, 82, 83, 84–87, 88–89, 91, 92

adaptationism: adaptation and mutation, interplay of 104; reproductive failure and 128–29, 133, 135–37, 141–42, 146–47, 150, 151

The Adapted Mind: Evolutionary Psychology and the Generation of Culture (Barkow, J. H., Cosmides, L. and Tooby, J., Eds.) 77–78

Adriaens, P. R. and DeBlock, A. 137, 138, 139

adulterous desires, narrative attraction of 15–16, 96–126; "antigravity" hypothesis 134; communicative failure 114–15, 116–17; consciousness, theories of 109; courtship and courtship rituals 110–12, 115, 123–24; DNA (deoxyribonucleic acid) 106, 108, 118; eggs as energy consuming 96; evolutionary biology, adultery and 109–10; evolutionary infidelity narrative 96–98, 103–8, 109, 112–13, 114, 117, 118–19, 120, 122–23, 124–25, 126n5; evolutionary psychology, discourse of 97, 98, 99, 100, 101, 103, 106, 108, 109, 110, 111, 112, 113, 116, 117–19, 122, 123–24, 125, 126n6; female sexuality 100, 139; female sexuality, representations of 104–5; fictional explorations of promiscuity and reproduction 108–12; gender, popular discourses of 97; gender, portrayal of characteristics 102–3;

gender differences 102, 104, 123, 131; genetic determinism 113; genetically controlled infidelity, romantic love and 112–17; "Homosexuality and Narrative" (Allen, D. W.) 121–22; honeybee mating 107; infidelity, maladaptiveness and 111; infidelity narrative 96–98, 103–8, 109, 112–13, 114, 115–16, 117, 118–19, 120, 122–23, 124–25, 126n5; male infidelity, genetically programmed 114–15; male sexuality 113; men's adulterous inclinations, folk wisdom of 98–99; monogamy 97, 104–5, 107, 134; morality, sexuality and evolution 99–100; narrative mutations 124–26; natural selection of genes 97–98, 101; neo-Darwinian "reproductive imperative" 96–97; parental investment, theory of 96–97, 99; "Parental Investment and Sexual Selection" (Trivers, R.) 96–97; polygamy 103–4; popular science texts 97, 98, 104, 118; primitive inclination toward "[u]tter licentiousness" 100; primitive societies, sexual selection in 100–101; promiscuity, adaptation and 107; promiscuity, infidelity and 97, 103, 104–5, 112, 122, 156; reproductive imperative 97–98, 105–6, 113, 118, 120, 123, 125, 126n5; romance, narrative structure of 98, 110–11, 115–16, 117, 119, 120, 121, 123–24, 125; sexual difference, primacy in Darwin's theory 102; sexual selection, gender characteristics and 102–3; sexual selection, natural selection and "sexual struggle" 99, 101–2; sexuality, popular discourses of 97; sexuality, sociobiological portrayal of 103;